Why Chimps
Can Read

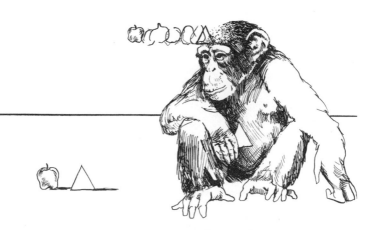

Why Chimps Can Read

ANN J. PREMACK

Drawings by Robert Schneider

HARPER & ROW, PUBLISHERS

New York, Hagerstown,

San Francisco, London

FIRST EDITION

Designed by Janice Willcocks Stern

Library of Congress Cataloging in Publication Data

Premack, Ann J
  Why chimps can read.

  Bibliography: p.
  Includes index.
  1. Chimpanzees—Psychology. 2. Animal communication. 3. Reading. I. Title.
QL737.P96P73   1976   156'.3   73–14283
ISBN 0–06–013389–9

76 77 78 79 10 9 8 7 6 5 4 3 2 1

to Jimmy James
my father

contents

illustrations

preface

In the course of preparing this book, I visited Dr. Roger Fouts in Norman, Oklahoma, and spent three eventful days observing Roger teach signs to several chimps. During that time, I had the opportunity to meet a few chimpanzees, including Lucy, who fed me, and Booie, who unlaced both my shoes and examined my toes inquisitively. I also had the privilege of visiting Harry F. Harlow's laboratories in Wisconsin, where Dr. Stephen Suomi provided an informative tour in the absence of Dr. Harlow. It is here that numerous clinical studies with rhesus monkeys have been performed—on mother love, isolation rearing, and depression—basic research in social learning. My thanks to both Dr. Suomi and Dr. Fouts for being generous with their time.

I worked with retarded children for some years, but left the field to raise my own family, and subsequently became involved with my husband's research. From 1954, when we spent a year at the original Yerkes Laboratories in Orange Park, Florida, until now I have been an active participant in all phases of the chimp language research; planning experiments, teaching the reading system to Sarah, and writing articles about the research in *Scientific American, National Geographic,* and *La Recherche.* I never dreamed the studies with chimpanzees would describe a circle, returning as a procedure to help retarded children. It is with these children that I have the strongest emotional ties, and I am grateful the chimpanzee research, devised for studying intelligence in the infrahuman, is providing both the idea of language, and the ability to speak, to hundreds of previously silent children.

My grateful thanks to Alpha, Ioni, Viki, Washoe, Sarah, and the many other chimpanzees who introduced life and humor to scientific studies. And my particular gratitude to all the experimentalists who devoted many years of their lives to the study of chimps. Their ingenious ideas have made this book possible. After more than fifty years of scientific inquiry, the bridge has been discovered where ape and human might meet—a historic event—exciting yet sobering, and worth examining carefully.

Ann J. Premack

Philadelphia, Pennsylvania
1975

1

the social

life of wild

chimpanzees

In the evening, just as the countryside begins to darken, chimpanzees prepare their sleeping nests, a chore which takes only a few minutes. The chimpanzees place leafy branches in layers to form a springy mattress, infants settle in the commodious beds their mothers build, and the African night grows silent except for the occasional cry of a restless creature.

In the early morning, the animals rise and relieve themselves over the sides of their nests. Some begin to eat ravenously, gobbling even the skin and pits of fruits. If nothing in the immediate vicinity is appetizing they climb down from their forest bedrooms and organize food-searching parties. Members of the band, about fifty in all, divide into smaller groups of both sexes, young and old, while mothers remain behind with their infants. As soon as a search party locates a preferred fruit, members call loudly and drum on trees to notify other chimpanzees, who hasten in the direction of the sounds: adult males, females with infants clinging to their underbellies, and young chimps.

Chimpanzees spend about six hours a day eating, and, after each leisurely meal, they relax. Young chimps scuffle with one another under the watchful eyes of their mothers, mothers groom and exercise their infants, and in the shade several males line up to copulate with a receptive female. A young chimp watches the proceedings closely, making a manual check now and then of the sexual connection. Down the road, two older males engage in termite fishing, using prepared sticks, while not far from them a young male tries to ant fish but, lacking patience, removes his stick too early from the tunnel, burrowed into the ant mound. Finding a mere three or four ants on the stick, he gives up and wanders over to a friend for a little afternoon roughhousing.

This account of a typical day in the life of a chimpanzee troop could not have been written fifteen years ago. Even as recently as the eighteenth century the orangutan was confused with the chimpanzee and regarded as a red variety; not until the beginning of the nineteenth century was the chimpanzee finally distinguished as a separate species. By the end of the century, four types of anthropoid apes had been identified: the gorilla, chimpanzee, orangutan, and gibbon. Up until twenty years ago, much of the information about the apes was based on myths, reports of natives, and stories told by hunters. But there were a few reliable reports mixed with the fanciful tales.

One of the early naturalists to take an interest in chimpanzees was Richard L. Garner, who tried to observe wild chimps from a cage he built in the French Congo almost eighty years ago. It was Garner who was confined to the cage, not the chimps, and he remained in his protective home most of the day for a period of four months, waiting to observe wild chimps as they roamed about the countryside. But they came nowhere near his camouflaged home. Undaunted, Garner adopted a few young chimps which natives had captured, and observed them instead.

From the natives came intriguing but fanciful stories. On

hearing chimpanzees built houses with roofs, Garner was so fascinated he looked everywhere for them, but despite his diligent search he never found a roofed house. The story that chimps constructed drums of earth and water was one Garner did believe, for he reported having seen such a drum himself, and with poetic detail he described the long, mysterious finger marks left by the chimp drum maker on the soil drum.

Several modern anthropologists have described noisy celebrations held among chimpanzee bands when groups joined together to celebrate the "harvest" of a favorite fruit. The forest reverberates with wild drumming and awesome party shouts, and the excitement grips even the humans who happen to be in the vicinity. Of all the stories Garner heard concerning wild chimps, that of the "kanjos" or carnivals is the only one verified by recent observations.

Though Garner was not a psychologist, he was interested in the intelligence of chimpanzees, concluding that the best way to measure an animal's ability to reason was to put the animal in a problem situation, then permit it to solve the problem with its own resources.

Some years after Garner's books were published, Robert Yerkes envisioned the idea of establishing a primate research center (the first ever built in this country), which soon became the famed Yerkes Laboratories of Primate Biology at Orange Park, Florida. There was little information on wild chimps to guide him in establishing a viable colony of captive animals, so he quickly sent Henry Nissen, a young psychologist, to conduct a field study of chimpanzees in French Guinea. Nissen's task was to observe chimps and capture a few specimens for the laboratories. In 1929, Nissen left New York for French Guinea, where he bought two infant chimps from natives, captured sixteen others, dispatching both jobs with ease; but observing wild chimpanzees was not quite as simple. Field techniques which were ordinarily used with most wild animals did not work with chimpanzees. Since chimps travel over a wide territory during the course of a day, sometimes not returning to a specific area for weeks, blinds were useless. Like Garner in his cage, a chimp observer might sit behind a blind for hours or weeks without seeing a single chimpanzee. Nissen's food lures were generally ignored by bands of chimps. Pressed for time, Nissen decided to trail the chimps by following their dropped fruit peelings, feces, and newly broken branches: a detective on the track of runaways. And when Nissen and his guides were lucky, they could follow the sounds of chimp cries and drumming. In the dark daytime forest, Nissen was unable to take any pictures, relying on his binoculars and notes instead. On rare occasions he watched chimps from as close as 15 to 20 feet, though the usual distance was 50 to 150 feet.

When his guides located the spot where chimps had settled for the night, Nissen got up early to observe the chimps at their morning activities. They rose with the sun, he noted, and ate lavishly in the mornings before leaving the nesting area. Curious about the construction of nests, Nissen and his guides examined several of them. In French Guinea, chimpanzee nests are 40 to 60 feet up, situated in the crotches of tree branches, looking like woven beds of soft leaves. Much to the surprise of all observers, nests were kept scrupulously clean; chimpanzees, even young ones, never soil their nests.

Though Nissen could not comment on the intelligence of the chimps, he did have opinions about their ability to communicate. The many vocalizations he heard from chimpanzees, Nissen decided, expressed individual body states and emotions only. By tactual cues, however, and by drumming on the base of tree trunks, chimpanzees exchanged information, he concluded.

Robert Yerkes was able to combine Nissen's field study with information offered him by Ms. Rosalie Abreu, a wealthy woman who kept a small colony of captive primates in her native Cuba. Ms. Abreu has the distinction of having been the first person in the western hemisphere to breed primates, including chimpanzees, in captivity. After she died in 1930, Dr. Yerkes was offered several chimpanzees from the Abreu collection, mainly females, a part of the original colony of animals at the famous Yerkes Laboratories, which produced an enormous amount of research over a period of forty years.

Yerkes also made young chimps available to outside researchers interested in comparing the child with the ape, as Winthrop and Louise Kellogg did when they raised the chimp Gua with their son Donald. Viki, who learned a few words of English under the tutelage of Cathy and Keith Hayes, was also from Yerkes Laboratories. While the institution was in an active phase, a number of scientists and experimentalists learned about chimps in an idyllic setting. Not only did humans familiarize themselves with apes, but chimps had to cope with humans in laboratories and in private homes.

There is a tendency to regard what is learned in the laboratory as somehow inferior to information gathered in the wild. But what we know about chimps today comes from many settings: from laboratories, from homes where chimps were raised from the wild, and on occasion from breeders and zoos. We can study man in many settings as well. Man can be studied as a simple hunter-gatherer far from civilization, in modern institutions such as prisons, or as he hustles doing a complex job in downtown Manhattan. Knowledge about chimpanzees comes legitimately from many spheres, and for almost forty years Yerkes was the citadel for scientists interested in man's closest relative.

But, when Yerkes established his laboratories, little was known of the natural life habits of chimps, and Henry Nissen's observations, regarded as a landmark study at the time, told us little about the social life of the elusive chimp. When we finally were able to read detailed descriptions of the life of the wild chimp, the Orange Park laboratories had closed for lack of funds, the grounds sadly deteriorated into a jungle.

Out in a beautiful but isolated country, the work of one Englishwoman stimulated an enormous surge of popular interest in the chimp. She was Jane van Lawick–Goodall, who observed chimps of the Gombe Stream Reserve in Tanzania. There were also Vernon and Frances Reynolds, who observed forest-living chimps by tracking them in the Budongo forest of western Uganda. In the Belgian Congo, Adriaan Kortlandt studied chimps by ensconcing himself on a plantation on the edge of a forest, where bananas and pawpaws lured marauding chimpanzees.

Since the Reynoldses were observing chimpanzees in a rain forest, they had no choice but to use tracking procedures in order to keep up with the animals. By contrast, Jane Goodall, working in a savannah-like reserve, had a rather clear view of chimps most of the time, and was able to spend her first three months studying the basic movement patterns of the colony from a distance. She moved closer to the hub of activity as the chimps were sensitized to her presence. At the end of fourteen months of patient watching, which brought her nearer to chimps than chimps had ever before permitted, she was amply rewarded; the chimpanzees continued their activities when she sat a mere thirty to fifty feet away. What we presently know of the fascinating and rich social life of chimpanzees is a composite tapestry of the observations of particularly the Reynoldses and Jane Goodall.

Basing a language on social behavior

Because we know little about either the vocalizations or the gestures used among wild chimpanzees, we must rely on their social behavior to tell us if they can be taught language. The method is indirect, but not without value.

There are about two dozen vocalizations, or calls, which have been identified among wild chimps, but virtually nothing is known about the meaning of the various calls. Henry Nissen, about fifty years ago, suggested the vocalizations expressed body states and emotions. Today this remains the accepted analysis of chimp calls. Chimps express fear of strangers, joy during meals, and satisfaction at being groomed by vocalizing in a variety of ways. Wild vervet monkeys, which are considered far less intelligent than chimps, have a repertoire of thirty-six calls, each unique. When Thomas T. Struhsaker studied communication among vervet monkeys in Kenya about ten years ago, he described the distinct calls made by vervets when they spotted three different types of predators: birds, mammals, and snakes. Upon seeing an Egyptian cobra, the monkeys would cluster in close formation about five feet away from the snake, giving the snake *chutter* call while staring as though hypnotized at the predator. The *rraup* call was sounded when eagles were about three hundred yards away, whereupon vervets within hearing distance took cover in the dense thickets. Sometimes the call was sounded for the wrong bird. When that happened, the vervets ran only part of the way, took a closer look, and returned to their former activities. The *uh* call alerted other vervets to mammalian predators such as hyenas or cheetahs, but the monkeys did not take flight unless the *nyow* call announced an unexpected movement of the mammal.

The calls of chimpanzees—the pants and low grunts made when chimps greet one another or eat preferred foods, the hoots at the approach of strangers—are not nearly as sophisticated as the vervet calls, which actually distinguish different predators by different vocalizations. The vervet calls, varying in sound and identifying different types of predators, almost have the status of words.

Calls are vocalizations considered by scientists to express emotion, not information, and are thought to escape uncontrollably when an animal experiences joy or fear. But it is entirely possible we might be missing something in accepting the traditional analysis of calls, considering how specific are some of the calls of vervets. Struhsaker worked very closely among the vervet monkeys, far more closely than a human can get to a wild chimp. Perhaps in the future a disguised human may impose his presence upon a troop of wild chimps and discover a semantic key to the calls of chimpanzees.

Though we cannot interpret what chimp calls mean, we can subject the social life of these animals in the wild to a more careful examination. It is in the social life of chimps that we find one of the most important requisites for language: the ability to make distinctions. For instance, chimpanzees make distinctions among their friends by greeting them in a variety of ways, depending on the depth of affection they feel. With some, they touch hands lightly; with others, only wild hugs or fondling of scrotums will do; they express everything from cool affection to passion. Though greeting is a simple behavior, chimps

"With some, they touch hands lightly . . ."

use a variety of physical expressions and gestures to show friendship for Sarah, coolness toward Peony, or enthusiasm for Walnut.

Wild chimpanzees also make distinctions in the act of giving. Not only can a mother chimp give food to a child, she can also give her infant on loan to another chimp for a short period of time. Jane Goodall's recent discovery of hunting among chimpanzees also documents that large males, after hunting and killing a young baboon or monkey, will distribute portions of the meat to every animal

"In species where giving occurs . . ."

which holds out its hand in request. In species where giving occurs only between mother and child, and where only food is given, the act may be an unconscious one, but among chimpanzees, where a variety of individuals give different objects to one another, giving probably is understood.

Suppose, for simplicity's sake, that language is no more than a system of labels which can be applied to living creatures, the actions they perform, places they go, objects they touch, and so on; then animals which can make distinctions among people, actions, locations, and objects should be able to learn a system of labels for these categories. If an animal cannot distinguish members of its own species, cannot tell a Peony from a Walnut, there is no point in providing labels, or names, for them. But if an animal demonstrates it can make a large number of distinctions, in a number of categories, we can teach it labels for what it knows. People teach various labels to their young: spoken and written words; and if children are deaf, signs; if blind, Braille. Chimps also can be taught "symbols" of various kinds as labels for other creatures, actions, objects, and locations.

Since chimpanzees can make distinctions by greeting different chimps in various ways, can differentiate between giving food and giving a young chimp to another for temporary care, we should be able to teach them labels for these social behaviors.

But the chimp is capable of far more complicated activities than those of greeting and giving, behaviors which involve only two chimps. The search for food, for instance, calls for the cooperation of most members of a troop. While infants and mothers remain in safety, the troop splits into smaller groups in search of food, and when a desirable fruit is discovered, other members of the group are called. Chimps eat more than thirty varieties of fruits, twenty types of leaves, six blossoms, a few types of monkeys, and a variety of seeds, stems, and bark in the Gombe Stream Reserve, according to Jane Goodall. Further, the animals prefer to eat their fruits at particular stages of ripeness. When the chimp troop splits up in the morning, each smaller group must include at least a few members who can distinguish those fruits which are edible from those which are not. Before the call is sounded to other chimpanzees to "Come and get it," a fruit's distinctive shape, color, aroma, and taste must be considered.

Chimpanzees also engage in a couple of sporting events—ant fishing and termite fishing, the latter being more popular. A chimp needs to avail itself of some essential basic information to be a successful termiter or ant fisher. Termites must be in a stage of readiness for flight or they will not climb onto the sticks which chimps insert into their mounds. The termite season lasts from October to January in the Gombe Stream, a period of short rains when the worker termites extend living passages up toward the surface of the earth as they prepare the migration path for the fertile winged termites, which live in the center of the mounds. Certain chimps seem to be aware of the general season when termites should be, so to speak, ripe. Jane Goodall describes how some chimps approach the termite hills at about this time, examining and scratching the mounds like professionals. They visit the hills several times in the weeks before termite migration, and when the hills are ready, the chimps prepare their sticks and begin to fish. For termite fishing, sticks are carefully stripped of leaves and measure a uniform twelve inches long.

For ant fishing, chimpanzees prepare and use sticks that vary from about two and a half to three feet in length. They poke the sticks into ant nests, keep them inserted for a period of time, then withdraw them to lick them clean of ants.

These simple recreational sports have complicated overtones, for, although most primates use their hands as tools, not all can use their hands to *prepare* tools. If chimps prepared only a specific tool for a specific activity, one might question whether the animal was fully aware

"Certain chimps seem to be aware of the general season . . ."

of what it was doing. But chimps prepare very differently for two kinds of fishing: long sticks are for ants, shorter ones are for termites. Because two types of fishing are clearly distinguished chimpanzees could be taught to apply different names to insects, types of sticks, the various types of mounds, and so on.

Most scientists agree that animals, especially apes, express themselves with calls and gestures. Whether the system expresses only emotional states, or whether the calls and gestures communicate information as well, is still argued. There are a few enthusiasts who believe wild chimps use language, but convincing others of this possibility is unlikely. Man regards language as his unique domain and guards his territory zealously, as well he might, for there is little evidence for language among wild chimps. So far, field observers still cannot identify the meanings for the simple calls and gestures of wild chimps.

Chimps are unable to produce human speech sounds, and we cannot decipher their calls and gestures accurately. The chimp, however, is closest to us of all apes in both intelligence and emotion and might be able to learn a simple language. If we invented such a language, consisting of about 150 labels, where each label was a "word" and words consisted of the names of chimps, people, actions, places, food, we could place these labels on a table, "write" to the chimpanzee, and the chimp could "read" and reply. We could, even with a simple vocabulary, increase our knowledge about one another: ask questions about chimp fishing preferences and procedures, "listen" to a chimp describe the flavor of termites, or request the taste of a jungle fruit. Further, one chimp could use the language to request grooming from another chimp; or ask humans for bananas, or receptive female chimps for an afternoon of copulation.

Picture two chimpanzees, one young, the other old, ant fishing in the bucolic setting of the Gombe Reserve. The older, a seasoned fisher, manages to retrieve stickfuls of ants from the mound. The younger grumbles because he can't trap a solitary insect. "No wonder," exclaims the older, "you're using a termite stick. Sticks for ant fishing are at least *this* long." But the young animal fails to retrieve any ants even though he has later prepared an appropriate stick, and asks longingly, "What do ants taste like?" "They're delicious, and if you want to catch lots of them, you have to keep your stick in the mound. If you remove it too quickly, the ants won't have a chance to climb on."

Simple language exchanges are already occurring in a

few laboratories which have language training programs for chimps. Because the social setting differs from that in the wild, the kinds of exchanges or "conversations" are mainly directed by people who teach chimps about a human environment. If the system, learned in a civilized world, were exported to the wild, chimps and people could, in theory, "discuss" the social life of the wild chimpanzee.

2

chimps raised in
human homes

The wild is an ideal place for learning about the complex social life of chimpanzees, but gives us little information about the individual animal. To study how a chimp matures, for instance, we must remove the youngster from its normal mother and provide a human foster parent. Because chimps are so like us, young chimps have little difficulty in forming close and loving bonds with their human foster parents. In fact, because a deep dependence of chimp on human does occur, it is possible for the human to teach young chimps a variety of subjects, including language. But this mutual attachment can lead to unexpected problems as well.

People who raise chimps have high expectations for them as they have for their own children, and when the chimps don't perform at these levels, the "parents" are often bitter. Several women who have raised chimps in their homes mention their eventual disenchantment with the limitations of their chimps. Men as a rule experiment in a laboratory setting and do not form an intense identification with the chimp. But an acquaintance of mine, a man who is presently raising a chimp in his home, does seem to be showing all the symptoms of a doting father. He claims his chimp can learn more than any other chimp and boasts he can keep the animal with him for many, many years. Not satisfied with his chimp's accomplishments to date, he predicts what the animal will do five, ten, fifteen years from now. Poor chimp child if it is not an overachiever! One of the painful aspects of raising a chimp as a child is that it is no longer possible to keep a chimp in a homelike setting after the animal has reached puberty, when it is about seven years old. Most chimps are then returned to the laboratories from which they came, for the humanizing period cannot last forever.

Aside from a human baby, I can think of no creature which can arouse stronger feelings of tenderness than an infant chimpanzee. It has huge round eyes and a delicate head and is far more alert than a human infant of the same age. When you pick up a young chimp, it encircles your body with its long trembling arms and legs, and the effect is devastating—you want to take it home! But raising a chimpanzee in the home is not in the least like having a pet; it is more like having a child—a child who can overpower you before it is ready for kindergarten.

Ioni and Rudy:
a comparison of child and chimp

Nadezhda Kohts, of the Darwinian Museum in Moscow, fully expected her home-reared chimp to acquire at least one word of Russian: "The infant chimpanzee constantly hears human vocalizations, responds correctly to spoken directions, uses his own natural sounds for expressing his emotions and acquires complex conditional reflexes for the mimetic expression of his desires. But never once has there been traced any evidence to the effect that the chimpanzee would try to imitate the human voice or master be it even the most elementary words of which he would be able to greatly facilitate intercourse with his master." Most of her observations of her chimp, Ioni, spanned the early years, 1913–1916, but her book, *Infant Ape and Human Child,* was written many years later, in 1935. Only after she gave birth to her son Rudy did she consider the possibility of comparing his development with Ioni's. And, by that time, of course, Ioni was no longer in the Kohts household.

Ms. Kohts received recognition in this country for her early studies on the vision of the chimpanzee. She was the first experimenter to use the match-to-sample technique with chimps, an infrequently used procedure because it is open to the criticism of providing unintentional hints to the animal. But she employed every safeguard and her technique was faultless in execution. To assess vision, Ms. Kohts placed a few samples of chromatic colors on a tray before Ioni. She held a specific color card in her hand, requiring Ioni to pick from his series on the tray the color that matched Ms. Kohts's sample. After deciding on a match, Ioni handed it to Ms. Kohts. He was able to distinguish colors of both halves of the spectrum and, surprisingly enough, picked the correct match for the sample whether he had two or twenty different color choices arrayed on his tray. In tests with two-dimensional shapes, none of which were colored, Ioni correctly distinguished and matched circles, ovals, rectangles, and half circles. He matched any number of pointed shapes, prisms with many surfaces, pyramids, cylinders, spheres. These elaborate tests gave incontrovertible evidence that the chimps' color and form vision is comparable to ours. And Robert Yerkes praised the research as "a major contribution to the comparative physiology of vision."

Ms. Kohts observed Ioni from the age of one and a half

to four, and ten years later she studied her son Rudy, from the time of his birth in 1925 until he reached the age of four. During these years, she kept a "mother's diary" on Rudy, but with no intention then of converting her copious notes into a comparative work as she later did do.

Unfortunately, *Infant Ape and Human Child* has not yet been translated into English, although the book includes a fairly complete English summary. I had been curious about it for some time, but could not find a copy, nor could I find anything except the most cursory mention of her work. I was surprised and delighted when David, my husband, came home one day with the set of books, a gift from Dr. Harry Harlow of the University of Wisconsin. Ms. Kohts had sent the books, the written work and its companion book of photographs to Dr. Harlow in admiration of his work with monkeys, and David in turn received them with Dr. Harlow's compliments. I could not have been more excited and went to work immediately, struggling to hone my four years of college Russian (now quite rusty) into a working language. From my efforts at translation, and from the English summary, a unique picture of the chimpanzee began to emerge.

A comparison of two individuals, one a chimp, the other a child, raised almost ten years apart was necessarily difficult, but Ms. Kohts gives the impression the two youngsters were raised at the same time, as if a time dislocation did not exist. Rudy and Ioni were alike in many ways. Both child and chimp "matched" their behaviors with the behaviors of the adults around them. We usually call this kind of matching imitation, and both youngsters imitated a number of adult household routines, including sweeping the floor, unlocking padlocks, hammering nails, using a stick to retrieve objects, turning on switches, unhooking hooks, and drawing patterns on paper with a pen or pencil.

If an essential object was not available when the youngsters were engaged in imitating, they found substitutes. For instance, when they had no pencil, each attempted to write with a nail, a stick, or even his own fingernail. In the absence of ink, they dipped their pens in milk, broth, or water, or even spit out some saliva as a substitute. They both enjoyed hammering, using their fists, stones, or any heavy object to drive a nail when a hammer was not available. This kind of substitution, also called "transfer" or "generalization," can be found in virtually all species, including the humble rat. But in chimp and child, transfer has a fascinating quality; and when a chimp learns a few "words," as some have recently done, the phenomenon is even more intriguing.

Both chimpanzee and boy understood the concept of similarity, and Ioni, partial to the color light blue, often selected blue objects to play with. At times, he gathered a series of small, round patterns into a pile. Rudy enjoyed noting similarities as well. When quite young he pointed to the legs of a horse he saw in a picture, then threw back his sheets, bared and demonstrated *his* legs. At the age of two, he pointed to the eyebrows of his doll; "brows-brows," he said and, pointing to each member of the family in turn, identified "mother-brows," "daddy-brows," and "uncle-brows." Recently, when a chimp named Sarah was taught a "written" language system, she learned to label "same" for similar objects, "different" for dissimilar ones, taking the basic concept of sameness in unexpected directions.

Rudy called a bat an "airplane" because each had wings, said "reindeer" to an interestingly shaped twig, and called the translucent stone on a ring a "bubble." He referred to large nails as "mother" and small ones as "children." Washoe, a chimp taught to label with signs, learned the sign for a flower, and later made the hand sign "flower" in the presence of pipe tobacco. Responsive to aromas, the chimp applied the same sign to both objects. This kind of generalization cannot be diagnosed an error, because, from the point of view of child and chimp, wings are more fantastic than animacy and aromas more enticing than the appearance of flowers. Further, the boy Rudy had not yet been provided with a name for a bat. Instead of inventing a new name, he applied an old name which seemed to fit. Washoe, the chimp, did quite the same thing by applying a known sign to a new item which had not yet been labeled. Both child and chimp used clever and original solutions when they found themselves without a name for an object.

For an animal to learn its own name, or be taught such pronouns as "I, me, mine," it must demonstrate a concept of "self." The chimp Ioni meticulously checked all parts of his body every morning, and if he noted any real or even imaginary defects such as scratches, specks of dirt, or soiled areas, he began to groom himself immediately, licking the sore place or removing the dirt. When Rudy was a year old, he also examined his own body with keen interest, checking his navel and fingering pimples or other intriguing spots. Both infants responded to their own image in a mirror first by smiling, then by touching the image. Both reached behind the mirror for the "rest" of the reflection. Finally, they spit at and made a series of unpleasant faces and noises at the mirror, ending by hitting the image aggressively. But, when Rudy was about two years old, he stopped attacking his reflection and began to show signs he recognized his image in the mirror. He brought his face near the mirror, kissed the image, and

"After the first twenty hours with the mirror . . ."

when asked, "Who's there?" said his name. Ioni never reached the stage of self-recognition.

Recent studies on mirror recognition by G. G. Gallup,

Jr., use full-length mirrors, which are left permanently in the cages of chimpanzees. After the first twenty hours with the mirror, the chimp shows signs he recognizes the reflection of his own person, for when a part of his body is painted which he cannot possibly see without a mirror, he uses the mirror as an aid to cleaning off the paint. Monkeys exposed to the same treatment never show these signs of image recognition—one difference of many separating the ape from the monkey. In brain capacity as well as in blood serology, the chimp is more like us than like the monkey. The chimp is a hominoid and belongs to the same superfamily as humans. The monkey does not.

Both Rudy and Ioni were fascinated by sticks of every sort. Whenever Ioni found one, he promptly began to dig in the ground. If he happened to be indoors he banged it against the floor, retrieved out-of-reach objects, or brandished it as a threatening weapon. Young Rudy used his sticks to draw signs on the ground, threw them over fences, chased dogs with them or used one as a bat to send objects flying in the air. He couldn't resist saying, "Nice stick," whenever he saw one lying on the road. He had, in fact, gathered a collection of sticks and relished presenting lectures on the particular function of each.

Like many scientists working with chimps, Ms. Kohts hoped to find much more in her chimpanzee than she did. Ultimately, Ioni failed her by not becoming human. She was annoyed: "And now, as I come to the end of my comparative study it seems as if the bridge by means of which I had all the time been endeavouring to span the gulf between ape and man was all gone to pieces. . . . Day in, day out, with infinite care and patience, steadily checking up the strength of each structural member did I construct this bridge. All the time it seemed to me that I was nearing the completion of a stupendous structure which was to link up the hitherto disjoined—and all of a sudden what came but complete collapse! I thought that I might take my two little ones each from his own side and bring them to the middle of the bridge so that after their long and arduous journey they might here at last shake hands." Though Ms. Kohts failed to make a child of her chimp, she described with great clarity how much less human is the chimpanzee than the child. Her careful and imaginative reports comparing the behavior of Ioni and Rudy continue to instruct us in the marvelous and unique nature of both chimp and child.

Alpha: an infant chimp matures

More than forty years ago, Dwina, one of the four original chimpanzees at Yerkes Laboratories, gave birth to a female chimp, which she promptly abandoned. Dwina did not neglect her child because of lack of motherliness but because she was ill. She died of puerperal fever two weeks after delivery. A couple of young psychologists engaged in research at Yerkes at the time, Carlyle and Marian Jacobsen, became the tiny chimp's foster parents, naming her Alpha because she was the first chimp born at the laboratories. Alpha spent a large part of the day, and frequently the night, with her substitute parents, but the Jacobsens made no attempt to humanize her in any way, either by teaching her our habits or by dressing her in clothes. Their monograph, published in 1932, was the first account of the physical maturation of a baby chimp raised in the absence of its natural mother.

Alpha was dependent in her first month, spending most of her time sleeping, and was active mainly during her feeding periods. In her second month, she began to investigate objects with her mouth, engaging in simple play and grasping. During her first three months, she was a busy thumbsucker, and in her fifth week she was sucking not only her thumb but her toe as well! In the seventh week, Alpha could reach and grasp her toes with good coordination.

She sometimes opposed her thumb and fingers to grasp objects, but this use was accidental for the most part. She used the squeeze grasp for medium or large-sized items, since the chimp's thumb is placed on its hand almost parallel to the fingers, but in picking up small objects, such as a sugar cube, she used her thumb and forefinger quite carefully. She manipulated objects just as normal children do. When given a toy, for instance, both child and chimp first look at it, then reach for, handle, and finally grasp it. Alpha explored most objects with her mouth, but when using her hand, she relied on her forefinger to examine most objects. Children often put objects in their mouths, but not to the extent Alpha did. She examined the nose and teeth of her human friends thoroughly, smacking her lips as she traced their facial contours with her forefinger.

By the time Alpha was three months old, she had already made enormous strides in her development. She had begun to crawl, sit, and even attempt some rudimentary walking. In the fifth month, she was climbing and jumping, engaging in play attack and a fair amount of threat behavior. Tests in the Gesell series showed Alpha far ahead of the normal infant in all locomotor activities, such as creeping, sitting, walking; and when Alpha was a year old, her physical maturity was equivalent to that of a two- or three-year-old child.

The Jacobsens observed a few of Alpha's attempts to communicate, though they confined the majority of their observations to the chimp's physical development. During the eight months in which they studied her, they never heard her babble, coo, or engage in any type of vocal play. In her first two months, the only vocalization from Alpha was a high-pitched scream of fear, but at two months she began to respond to the Jacobsens' "uu-uu" chimp bark by turning her head in their direction, protruding her lips and answering with her own call. She showed considerable and approving interest in human faces in her fourth month and produced her soft guttural "uu-uu" at the approach of food. At the Yerkes Labs, she turned her head in the direction of the barking sounds of older chimps, responding to their calls with hers. Later, she used to bark whenever she was excited, as when she walked on two feet, when humans approached, or when her favorite food appeared.

When Alpha returned to Yerkes at the end of the year, she met her roommate Bula, an infant chimp from the Rosalie Abreu colony in Havana. Alpha proved herself a tyrant. She promptly snatched Bula's chimcracker, then chased her into the upper portion of the cage. Whenever Bula attempted to get down, Alpha raised her long index finger, pointing in the direction of the upper cage, a gesture that sent Bula scurrying in that direction again. Alpha clearly relished her power and dominance. She continued aggressive, Bula submissive. Bula did take a few liberties during this trying period, sometimes giving Alpha's hair a few nasty tugs when Alpha lay sleeping and it seemed safe. After several months, Bula came to challenge Alpha's authority and the two animals thereupon became mutually dependent and friendly.

Alpha provided a source of physiological information about the maturing chimpanzee: information on the size of its brain, length of its bones, teething development, and so on. The chimp seems to develop in quite the same pattern as we do, but the process moves far more rapidly in the chimp.

Since the Jacobsens limited their studies to the physical development of the chimp, we needed to look elsewhere for information about the social life of the young chimp living among people.

Gua and Donald: raising a chimp with a child

The vacuum of knowledge was soon partially filled in 1931 when Winthrop and Louise Kellogg adopted a seven-and-a-half-month-old female chimp and raised her as a virtual sister of their ten-month-old son Donald. The meeting between Gua and Donald, described by psychologist Winthrop Kellogg in his book *The Ape and the Child,* was gentle and charming. Donald sat within his play pen, while Gua, the chimpanzee, was brought in and placed on the floor just outside the pen. Each youngster was immediately fascinated with the other. The chimp inserted her long arm through the rungs of the play pen, reached for and held Donald's hand. She patted Donald gently on the tummy and Donald reciprocated by grabbing a fistful of Gua's hair. Suddenly the seated Gua lost her balance and fell over on her back. Seconds later, Donald fell over as well and broke into tears.

After the two had been living together for a few months, their interest in playing with one another was so intense that the one could not be induced to eat a meal while the other was playing. Gua ate the same strained baby foods Donald received, but she supplemented this basic diet with fresh vegetables she raided from the pantry. Outdoors, she ate flowers, leaves, and the bark from young saplings. She drank a great deal of water, particularly in the summer months, since she received very little fruit, the mainstay in the diet of the wild chimpanzee. One of Gua's favorite outdoor pastimes was ant tracking. She would follow a single ant for a considerable distance, then watch the activity around the ant colony for several minutes, though Gua was never observed to eat ants.

The two playmates loved to sit in packing boxes, bushel baskets, and even large pots and pans. They adorned themselves with blankets or clothing, which they draped over their shoulders and dragged about. Gua would frequently wrap herself in Spanish moss, or place foliage branches on her body. When Gua was eleven months old, Donald thirteen and a half months, they enjoyed playing the game of "giving." Dr. Kellogg would take an object and say, "Now I give this to Gua," "Now Gua gives it to Donald," "Donald gives it to Daddy," and so on. The two youngsters played the game for long periods of time, always identifying the recipient correctly. Later, they rolled a ball to one another, according to the same directions.

Both were tested on a series of comprehension tests designed by Arnold Gesell. Donald performed adequately, but there is evidence that, just as the environment encouraged Gua's human development, it somewhat hampered Donald's. He was slower than normal in his language development. Though the causes may be multiple, Gua's silence as a companion and the absence of talking children as playmates for Donald may have retarded his speech, temporarily at least. Gua understood a number of verbal commands, some salutations, and a few instructions. Her language consisted mainly of a number of physical expressions which both the Kelloggs were able to interpret. When Gua was hungry, for instance, she promptly climbed into her high chair. She protruded her rather large lips toward her cup when she wanted more milk, subsequently pushed the cup away or turned her back when she'd had enough. At the completion of her meal, she removed her bib and climbed out of her high chair. She tripped over to the five-gallon water bottle and licked it to indicate her thirst; and when she wanted to "play," she would stand in front of her chosen partner, grab his hands in hers, and take a few sideways steps to show what she wanted. Indicating her toilet needs made up a large portion of her daily communication, and although she used physical gestures at these times, often her requests were accompanied by vocalizations.

What is remarkable in this account of a home-raised chimp is the relative ease with which the young chimp can adjust to the social life of a human family. Gua was intensely loyal to her family, became antagonistic to visitors who got into arguments with Winthrop, and clung to both Louise and Winthrop quite as she would have to her natural mother in the wild. While she could carry out a few simple spoken commands, this ability was less consequential than the fact that she could, using a series of gestures and activities, "tell" her human family what she wanted. After remaining with the Kelloggs for about a year, Gua returned to Orange Park and her chimp colony at Yerkes.

Viki: concepts which underlie language

Cathy, a graduate in journalism, and Keith Hayes, an experimental psychologist, were convinced that chimpanzees were sufficiently like man in intelligence to be able to learn to speak. Chimps did not use language in the wild, they reasoned, because there was neither incentive nor a language environment. If the chimpanzee was offered all the advantages of the human child, it could learn spoken English. The project seemed far-fetched to most people they knew, but in Orange Park, where chimp lore and life exuded from the earth, the effort seemed a reasonable step. People intensely interested in chimps simply found their curiosity whetted by the evidence of Gua's ability to communicate.

As Cathy points out in her book *The Ape in Our House,* Keith directed the research while she carried it out. But Cathy did more, for her popular account of life with Viki became a book much enjoyed by chimp lovers both in this country and abroad. However, Viki did *not* learn to talk. After many years of effort, after the Hayeses used all the learning and speech-therapy techniques available, Viki learned to say three words, "mama," "papa," and "cup," but could produce these sounds only with the greatest difficulty. Further she never used "mama" and "papa" to refer to Cathy and Keith, but used them to name any objects she wanted as toys or food. She did say the word "cup" when she wanted a drink, however.

That the chimpanzee was not able to speak was a disappointment for the Hayeses, and particularly for Cathy, who by now had developed a strong attachment to Viki. Viki was, for years, Cathy's child. She devoted many hours to educating Viki, testing Viki, caring for Viki, and being thoroughly frustrated by her. In many ways, Viki was very like a human child in the early years of her life, but some differences placed her at a disadvantage in the human home. She was extremely hyperactive, for one thing. For another, she could neither comprehend nor produce words spontaneously and failed any tests in which she had to comprehend and follow spoken instructions. "In the beginning we had argued that apes do not learn to talk because they are raised under such distressing circumstances. We pointed out that children in institutions very often have speech handicaps. We said that if an ape had a proper upbringing, it might learn to speak spontaneously. But we were wrong. You can dress an ape in the finest of finery, buy it a tricycle and kiss it to death—but it will not learn to talk," Cathy wrote, admitting their defeat.

But, when Viki was about four years old, Cathy began a series of tests loosely described as measuring higher mental functions. The chimp's performance on this battery of tests was compared to that of children in Viki's own age range. To see if Viki could match numbers, Cathy gave her pairs of large cards on which were painted different numbers of dots of various sizes. No cards were ever alike, even when they contained the same number of dots. If the number of dots was the same, their size and location on the card was entirely different. It was impossible for Viki to match the numbers of dots on the basis of similarity; she had to look for actual numbers of dots. When she was shown a card with three dots, these might be arranged in a diagonal line, with two large dots and a tiny dot. Then she was shown another card, perhaps with three dots as well, but they formed a triangle of tiny dots. If she matched the cards, she was able to match numbers.

Viki performed well on these tests only when the two cards were sharply different in the numbers of dots they contained. When her choices showed one versus four, or three versus six dots, for instance, Viki could pick the choice card that matched the sample card. But when the choices were three versus four or four versus five, Viki became confused and matched on a chance basis. She performed nicely with the lower-numbered cards, such as one versus two, and two versus three, however. Children tested on number matching, all of whom were about three and a half, performed the problem very much as Viki did. Two older children, four and a half and five respectively, were able to count and easily noted numerical differences on the cards no matter how slight.

When Viki was five years old, Keith and Cathy devised some novel tests to see if she could make certain kinds of conceptual distinctions. A small but growing number of psychologists are now emphasizing the importance of conceptual distinctions as a prerequisite to language. Though Viki could not speak, the Hayeses were sure she could make such distinctions without using language.

First, they wanted to find out if she could distinguish the two categories "animate" and "inanimate." They assembled a series of colored pictures, forty-one pairs in all, twenty-one representing animate objects, the rest representing inanimate ones. The animate cards included pictures of people and other mammals, birds, a few insects, and a snake. The cards in the inanimate group were mostly pictures of furniture, but there were also pictures of a car and a clock, both of which suggest movement. On the first trial in this test, Cathy placed a clear sample of each category in front of Viki, and Viki was required to place each card in her pile in the appropriate category, animate or inanimate. To perform the test correctly Viki had to use something like the concept "animate" in sorting her pictures. She scored 85 percent on this problem, a B+. Of the six errors she made, three were pictures of insects, which she judged inanimate, arousing our sympathy rather than surprise.

To see if Viki could discriminate "male" and "female," Cathy showed Viki sixty-five pairs of pictures of humans of both sexes. All were fully clothed, though some wore shorts or slacks. There were pictures of young girls and boys. Fifteen pairs showed only the head portion, while a few emphasized the facial features only. Viki chose correctly on only 67 percent of the trials. On a test of color distinction, Viki identified the colors red and green with an accuracy of 74 percent. Her score was extremely high, 89 percent, in discriminating between pictures of children and adults. When Viki's performance was compared to that of children in her age group, all youngsters performed at similar levels of accuracy.

On the male-female distinction, two of the three children surpassed Viki's performance, but the third child complained he could not do the problem. Cathy later found out the child did not understand the concept of sex. Many of the children relied on concepts somewhat different from those of the adults who arranged the tests. For instance, though the testers considered "adults" and "children" the category names, children categorized the groups as "little boys, little girls," and "mamas and papas," which could be subsumed, of course, under the broader concepts.

Viki was able to sort items quite expertly on the basis of a single dimension, such as form. She also showed considerable flexibility in sorting the objects on the basis of color. She sorted buttons, first by color (green versus red), then by size (large versus small), and finally by form (square versus circular). Viki changed from one way of sorting to another quite spontaneously; she needed no instruction to note that her pile of buttons differed in three basic dimensions: color, size, and shape. Sarah, the chimp I taught, went even further. She learned to label the three categories and wrote, in her language system: "banana" "is" "yellow," "apple" "is" "red," "red" "is" "color." She also distinguished "small" from "large," "round" from "square." As long as the chimp can make distinctions on a perceptual basis, it can apply labels to them. When Viki was given a playful test to find out how she viewed herself, she correctly sorted all the animal pictures together, including a photo of Bokar, her natural father, but she dropped the picture of herself in the human group, consisting of such personalities as Franklin Roosevelt, Joe DiMaggio and Milton Eisenhower!

Chimpanzees and people:
made for each other?

Many people regard baby chimps as substitute children. Cathy Hayes mentions her discomfort on seeing Viki after an absence of five days. Cathy managed a short vacation when Viki was about two, but was homesick for her "child" after the few days' separation and wrote: "I knew that this was my baby by her response to me . . . but someone had dressed my baby in a monkey costume."

A completely practical problem in raising a chimp is coping with toilet training. The wild chimp learns to be fastidious, but the home-raised chimp is an indoor animal and must be potty-trained. Probably the easiest way to handle a home-raised chimp is to diaper it, which the Jacobsens did with Alpha. When the animal gets older, however, diaper changing is a headache, as chimps have an active elimination system. Cathy Hayes had only moderate success with Viki, but the Kelloggs' Gua trained fairly well, perhaps because she and Donald were potty-trained together and found the sessions enjoyable.

How we respond to chimps is largely determined by the appearance and behavior of the chimp itself. The baby chimp is easy to love; not so the mature animal. Many feel threatened by the presence of the strong, hairy chimp, and retaliate by teasing the caged animals and trying to provoke them. Chimps reach puberty at about the age of seven, and it is rarely possible to retain a close working relation after that time. When chimps are younger they have temper tantrums and flail about when they make errors on problems. Though they can administer a bruise or two in the early years, the same tantrum can result in a serious hurt after the animal reaches puberty. It is not so much a matter of the animal's now becoming dangerous, but rather of its having become so strong. The same flailing arm, endearing in the infant chimp, now has all the qualities of a battering ram, and will send the poor experimenter reeling.

Cathy Hayes methodically removed all decorative touches in her home as Viki grew up. She stripped the house of all drapes finally, after having first reduced them from floor length, to window sill and half-window length. Her house was soon bleak enough for a visitor to ask as she gestured toward the empty living room, "at the clamped-down lamp, and the mere fringes of drapes," " 'Do you live here too . . . or is this just where you keep the ape?' " Some of the problems in working with chimps can be more subtle, as in my case. A few years ago I worked on a language project with David for a short period of time. Sarah, the chimp being taught "words" in a school-type setting, was a very bright animal, but I sensed I was working with a retarded child. I knew I could teach Sarah if only I had the patience and a good instructional program. Many days, when Sarah and I worked, I felt far removed from the rest of the world, knit together with Sarah in a secret world of our own. It was a strange, mysterious experience. Then, one day, when Sarah made many errors, I became enormously irritated with her and called her an "idiot."

Suddenly a flood of memories rushed into my mind as we sat in the dank laboratory. I saw myself seated at a table in my father's office at our house in Shanghai, trying painstakingly, at the age of about ten, to teach my sister of nine how to read. The book was elementary, but she barely knew the alphabet. I was furious with her for making errors, for not sharing the reading I loved, for being so slow, for not being normal. She was my retarded sister, recalled to memory by a chimp named Sarah. Suddenly it became clear why this project was so important to me. I clearly interpreted teaching language to Sarah as another chance to teach my retarded sister to read. And I hoped not to fail with this "sister" as I had failed before. For me, the chimp was not a substitute baby or a primitive human, but a retarded child.

How like humans are chimps? The study of individual chimps raised in a human environment clarifies some points of similarity: young chimps are like children in their physical development, sitting up, standing and running in a similar developmental pattern. In certain types of simple mental tests, young chimps perform about as well as young children. Though few scientists have done comparative studies, these did provide basic information which has been valuable to subsequent researchers.

But a chimp raised in a human home cannot act out a normal social life, nor can the chimp, after puberty, remain in a home. These animals are eventually transferred to zoos or laboratories, where the facilities are geared to adolescent and adult chimps. A laboratory does not stimulate the kind of personal involvement with the chimp that is found in a home. In the lab, chimps are not encouraged to behave like people, but like themselves. As a rule, an experimenter arranges a test, then brings a chimp or a group of chimps into the situation and observes what happens. As much as possible the experimenter excludes himself, not imposing human solutions on the chimp. In this situation chimps can behave much more naturally, as they might in the wild.

3

innovation and learning

Many anthropologists suggest language cannot be found in a species without culture. At a humble level, culture means "handing down," from one generation to the next, both the traditional and the innovative ways of a group. Parents teach their children to cook, prepare tools, plant gardens, marry, bear children, in the traditional ways, as a rule. Virtually all cultural knowledge can be handed down without language, since every animal can learn the activities of the group by simple observation. When language developed as a cultural innovation we abandoned learning by observation in favor of learning by verbal instruction. Listening and reading became popular. Instead of learning to cook only by watching mother at the stove, the child could learn more efficiently by way of language instruction. TV cooking shows retain observational examples to enrich verbal instructions, which may account for their popularity. Generally, language has replaced the gracious art of learning by observation with that of learning by verbal instruction and correction.

Geza Teleki, a student with Jane Goodall, in a *Scientific American* article, reports that about ten years ago only older males directed hunting groups, but now many more and younger chimps engage in the predatory tradition with greater frequency. They seek out, pursue, kill, and share small prey with other hunt members. The cultural practice of hunting has spread to younger members of chimp groups in the last twenty years. Nest-building, a common activity among chimpanzees today, was probably innovated hundreds of years ago by a single chimpanzee, then spread.

It is possible to imagine that day in the distant past when a young chimp found a spot of luxuriant foliage in the crotch of some tree branches, and began to weave leaves and stems into an oval nest. Though the bed was simple it secured the sleeping chimp through a long night. If the nest had failed to hold the sleeping inventor, the results would have been disastrous, plunging him forty feet to the earth. But it held. And others, seeing how comfortably the innovator slept, observed the technique of construction. In time, every grown chimp could build its own nest, while infants cuddled in with their mothers at night after observing their mothers prepare the bed. In the daytime, safely on the ground, they played at nest building with leaves and twigs, and later, they too constructed oval nests in the tallest of trees. But this is a speculative story, whereas the true account of an innovation and its "handing down" among a group of island monkeys puts the fiction to shame.

Innovations among primates

In 1962, a group of researchers from the Japan Monkey Center scattered sweet potatoes on the sandy beach of Koshima, a small island off Japan. Living on the island and foraging exclusively in the mountain areas was a troop of Japanese macaques, a species of Old World monkey with a very short external tail. The animals had never wandered on the sandy beach, nor had they ever approached the sea, but not long after the sweet potatoes were scattered on the beach, the macaque group began to abandon their old forest home and to forage among the potatoes. About a year after the artificial feeding program started, a two-year-old female, named Imo by the researchers, carried her sweet potato to the edge of a brook, dipped the potato in the water and washed off the sand particles. Most of the macaques continued to remove the offending sand particles by dry-brushing the potatoes with their hands, but in the ensuing years many of the animals began to wash their potatoes. Gradually potato washing shifted to the nearby sea rather than the brook.

The adoption of this new technique was slow. At the end of the first year, only four of the sixty animals followed Imo's example, the younger animals being the first to adopt the new trend. Five years later almost all the younger animals washed their potatoes, while few of the older animals adopted the custom. The invention and adoption of potato washing produced radical changes in the living patterns of the Koshima macaques. Instead of living in the forest, the macaques now became sea creatures. Earlier generations of the Koshima macaques may never have entered the water, but now infants were taken to the water, clinging to their mothers' bellies while potatoes were washed. At the age of six months, the youngsters began to pick up pieces of potato that were dropped, diving underwater to rescue the tidbits. By the time the youngsters were two and a half years old, they had adopted and perfected the practice of washing potatoes, members of the new generations both explored and played in the water, and juveniles learned to swim. This study permits us to see in clear detail how an innovation begins, is adopted by other members, and finally changes life patterns of all members of a group. It is a simple culture pattern which can persist for many generations.

The hamadryas baboon, which belongs to the family of Old World monkeys, not to the higher family of hominoid, to which both chimpanzee and man belong, lives in the semi-desert area of eastern Ethiopia and searches for a daily source of water. It was forced to be innovative, because finding a source of clean drinking water was often difficult. The hamadryas travel in arid, treeless country, a large group of females and young surrounded by larger males in a protective ring as they search for food and water. In this bleak environment, it is particularly hard to find water during the hot months, when the ponds in Ethiopia are greenish, and the water is tepid and unpalatable. The hamadryas baboons have invented a simple but ingenious method for obtaining filtered, cool water. The older male baboons dig holes around the periphery of the pond—some almost a foot in depth—and soon clear, sweet, filtered water rises in the holes, permitting the animals a cool drink on a parched Ethiopian day. This clever solution to the problem of polluted water is observed and repeated by new generations of baboons in a traditional pattern.

The technique of potato washing which changed the life of the macaques may be a simpler innovation than digging wells or preparing sticks for fishing, but the effect is the same—the innovation is disseminated among members of the social group and is subsequently adopted by the new generation, changing the group's traditional life in various degrees.

Young chimps observe ant and termite fishing, try the activity, make mistakes and later try again. No one corrects their errors or tries deliberately to provide them a proper example. Infants play at nest building during the day without interference from their mothers. Most of the behaviors, at one time possibly innovated by a single animal, are adopted by chimpanzees through observation. It is the same way normal children acquire their first language, not with deliberate instruction but by simple observation. But no human has had the privilege of seeing a chimp innovate in the wild; the birth of a new behavior has been hidden in the soft shadows of the forest. How and when nest building appeared, for instance, which chimps adopted the practice, or how this activity may have changed their social life is a mystery.

The experiments of Wolfgang Köhler, however, provided considerable information about innovation among chimps by demonstrating the "creative" manner in which laboratory chimps solved certain problems. They give us no information on the spread of the innovation to other chimps, nor do they offer any clues about deliberate instruction in teaching new techniques, but the studies are valuable for showing the origins of innovative solution to problems and for recognizing that chimps vary in creative ability. Some were endlessly imaginative and others had all they could do just to cope with the daily program.

Köhler studied a small group of chimps at the Anthropoid Station in Tenerife, in the Canary Islands. Completed almost sixty years ago, during the First World War, about the time Nadezhda Kohts was raising her chimp, Ioni, these notable experiments have had considerable influence on subsequent students of primate behavior.

Some of the daily activities Köhler noted among his chimps at Tenerife in 1913 were similar to those discovered only many years later among chimps in the wild. Given sticks and straws to play with, the laboratory chimps would use them for capturing ants. They placed the sticks outside their cages, gave the ants some time to climb, then licked their straws and twigs clean. It was fifty years later that Jane Goodall noted this use of tools among wild chimpanzees. On occasion, the animals used the sticks for digging in the ground, quite as Ioni often did. If the ground was firm, the chimp pressed the sole of its foot near the bottom of the stick like a shovel, driving it into the ground with considerable force. In the course of desultory digging, some of the animals discovered a few tasty roots and soon a large group of chimps took up digging.

Köhler arranged for each chimp to have a supply of straw and old rags and a collection of odds and ends in its home cage. The chimps wove nests out of this material and one chimp, Tschego, built very remarkable nests. If she found straw heaped on her sleeping board, she sat on it, bent a handful from the edge toward the interior, placed her foot on the twisted end, then worked around the periphery until she had formed a nest with a neat edge. Most of the animals built day nests of a variety of materials, such as straw, grass, branches, rags, and ropes. The chimp's main purpose seemed to be to build a ring around itself. If there was not sufficient material to fill a generous nest, the chimp sat, looking forlorn, in a rather skimpy circle.

Köhler was particularly interested in how learning occurred, concluding that if it occurred at all it took place in one trial only. He did not hold the view, popular in his day, that learning was a gradual process. The popular view could be explained, Köhler thought, by examining the experimental design of most learning studies—they *fostered* gradual learning. An animal in a box with a lever investigated the surroundings until it accidentally struck the lever. When it did, the animal received a reward, a pellet in a dispenser. After a few more accidents, the animal intentionally pressed the lever for the pellets. Köhler did not consider this procedure a good test of an animal's ability to examine its environment or to solve problems.

Rather than keeping food hidden, as in the lever box, Köhler made food highly visible at the outset of his experiments. Completely within view, the food was nevertheless out of reach. It lay beyond the bars of the cage, far too distant to be reached with even a long chimpanzee arm; or it was suspended from a height that exceeded jumping range. In the lever box, potential tools were visible but food was not. In the Köhler situation, the tools for obtaining the food were barely visible, but the food was in plain sight. The tools were within the animal's visual field, but only subtly so, and were sometimes mingled unobtrusively among the other odds and ends in the compound or cage. But the food was very visible and available to the chimp if it could find a way to obtain it. Köhler then proceeded to prove his point, that learning occurs in a sudden, insightful, and spontaneous fashion. He used seven chimpanzees in this series of experiments.

In the majority of the studies, a favored food, usually some bananas, was placed out of the animal's reach in either a vertical or horizontal plane. If the food lay beyond the bars of the cage, and well beyond arm's reach, the chimp had to find a way to "extend" its arm in order to reach the food. Most often, it used sticks to solve the problem, but it manipulated other items innovatively as well. If the food was out of reach on a vertical plane, and reaching and jumping could not bridge the distance, the chimp could employ a box or boxes to "extend" its height. The chimps invented reasonable but unexpected and amusing alternatives.

When Tschego, an older chimp, first saw some food lying outside her cage, she tried to reach the fruit with her hand, but failed. She lay down for a short nap, a bit frustrated, then tried again. She failed once more. After a half hour of wasteful effort, she looked around and spotted a set of sticks. They had been, of course, lying close to the cage, in plain view, when the experiment began. Suddenly Tschego leaped to her feet, seized a stick and pulled the bananas toward her. She used a raking technique, switching her working arms frequently as she tired, and moved with feverish haste.

After the chimpanzees had a few experiences in using a stick to retrieve food beyond the cage, they learned to be innovative even when a stick was not at hand. They learned to use any item that could extend the reach of the arm. Once again the phenomenon of transfer shows up. As with Ioni and Rudy, who substituted objects for missing objects—water for ink, a stone for a hammer in pounding a nail—these older chimps used rags, straws, even a tin drinking bowl as substitutes. One animal literally "beat" the fruit toward the cage with rags, the technique working very well, much to Köhler's surprise. Some of the objects tried were ineffective—too short, too soft, or too weak. Köhler did not observe whether the chimpanzees used alternatives in a systematic way. They tried every available substitute for the stick. But we do not know if they might have been more circumspect later, choosing only those objects that could work, and eliminating those which were impractical. If they did, it would be nice evidence of self-education or self-correction in learning.

All six of the animals were brought into a room with completely smooth walls and a roof too high to reach even in a leap. Köhler had earlier hung fruit from the ceiling, and in the middle of the room, plainly visible, stood a large wooden box. The fruit was nailed to the ceiling in a corner of the room some distance from the box. Each chimp tried first to reach the fruit with a powerful leap. Suddenly, Sultan, a young male chimp, tipped the box toward the fruit, climbed on it, sprang toward the food with all his strength and tore it loose from the nail. None of the other chimps had noted the connection between the box and the fruit.

Sultan was then placed alone in a different room, where the fruit was suspended at a much greater height than before, with both sticks and boxes available. At first, he tried to knock the fruit down with a stick, but he failed to reach it and succeeded only in tiring himself out. As he rested for a few minutes, his glance fell on a group of boxes. He immediately pulled one of them just under the fruit and grasped his prize. The next day Köhler took away all the sticks, leaving only boxes in the cage, but Sultan was unable to construct a solution. He tried piling the boxes on one another but he balanced them incorrectly and they tumbled over. Sultan's temper grew quite vile. In the middle of a paroxysm of fury, huddled in a corner, Sultan saw one of the caretakers walk directly beneath the suspended fruit in the experimental room. Sultan sped toward him, took his hand, pulled him toward the fruit and tried to climb onto his shoulder. The caretaker tried to free himself at first, but Sultan persisted, and in a few seconds he had sprung onto the caretaker's shoulder and reached the suspended fruit! Later on, this solution was preferred to the use of boxes, and soon all the chimpanzees were struggling to climb onto one another to reach the fruit. As each chimp struggled to climb on another, each also scrambled desperately to avoid being the bottom rung of the ladder for another chimp. There was no question that the chimps preferred to *use* someone else as a "box" than to build a construction of real boxes.

Köhler was curious to see if the chimps would pile boxes to reach suspended food. Not all of the chimps were able to do so. After some practice, Sultan and Grande, a female, reached the fruit by stacking and using two boxes, but these were stacked in a precarious and sloppy fashion. Sultan later succeeded in constructing a three-box structure to reach food, while Grande, a patient and careful animal, built a four-story structure. Chica, a young female, built a very tidy three-box building, Rana, another female, seldom built more than two-box constructions, and Tercera and Tschego made futile attempts, while Konsul, the other male, never built anything at all.

Köhler then gave a group of chimpanzees several opportunities to build cooperatively. Seeing the food aloft, the animals immediately began to look around for appropriate tools. But the chimps did not build cooperatively. Each chimp placed its own box in position, then reached for the next-closest box, which usually belonged to someone else. This would lead to a fight. When a sufficiently tall box building was finally built virtually by accident, the chimps promptly destroyed it in the struggle to be the first to mount it. Tercera and Konsul did not take part in the building operation but sat on the outskirts watching. When the required structure was almost complete, the two rascals crept up furtively and gave the mass of boxes and climbing bodies such a vigorous push that everything tumbled to the ground. The culprits then retreated to a safe place.

The everyday handling of objects by chimpanzees probably comes under the heading of "play." When the Tenerife chimps discovered a new method for using a tool in the experimental situation they quickly incorporated it into their "free play." Conversely, many of the activities used in play were put to work to solve the practical problems chimps faced in the experimental studies. Jumping with a pole, for instance, a form of play invented by Sultan, was quickly imitated by the other chimps. An animal placed a long stick, a pole, or a board either upright or at a slight angle to the ground, climbed up the stick as speedily as possible with both hands and feet, then fell with the stick or managed to jump off just as it was falling over. Sometimes an animal landed on the ground; at other times it flew into the trees or to other high spots. Later the chimps began to use the pole as a tool rather than a plaything. When a piece of fruit was hung beyond the animal's reach, the chimp first tried to leap straight from the ground, but when the attempt failed, it seized a nearby pole, settled it into the ground beneath the hanging prize, climbed up and wrenched the food loose.

Most of the chimps chewed on wood in play. If Grande wanted to make contact with a friend through the bars of her cage, she bit a board in pieces and used a small splinter section to poke her friend. Sultan sharpened a piece of wood with his teeth to poke into the keyhole of his cage. His effort did not work, but the spirit in which he attempted to escape was innovatively sound. Köhler decided to see if Sultan would "make" his own tools in order to reach food. He gave Sultan a stick with a rather large opening at one end and a narrow wooden board which had to be split to fit in the aperture of the stick to lengthen it. Initially, Sultan tried to fit the whole board into the aperture, then bit the board and broke off a long splinter, which was, unfortunately, still too thick to fit the stick's aperture.

Generally the stick served as a sort of all-purpose tool in the hands of the chimpanzees on Tenerife. The functions the stick could serve were transformed from day to day. Any attractive object which lay beyond the chimps' cage was reached with sticks, wires, or straw if it lay beyond arm's reach. When the rainy season ended and the favorite edible bushes were in bloom, the chimps pushed their sticks through the caging and forced the bushes against the cage, bringing delectable leaves for their eating enjoyment. Such occupation and preoccupation with sticks is familiar—Rudy and Ioni shared it and wild chimps strip branches for ant and termite fishing.

The Köhler chimpanzees solved their problems innovatively. It seems likely the more advanced species innovate frequently, making the cultures complex—there is more to teach and learn.

When a chimp becomes part of human culture, whether in a home environment or in a scientific laboratory, it becomes a victim of human ways: a victim of a culture where language is a teaching device and an uneducated chimp is a challenge; a culture where pedagogy is an obsession— where humans want, in a sense, to provide the chimp its missing cultural innovation, language.

4

In the wild, chimpanzees communicate with a number of gestures and calls. Chimps scream when in distress or when fighting, when the weather is rainy, or food is unpalatable. When they groom one another or eat a preferred food they grunt their pleasure, and they exchange low grunts during greetings, while they kiss, hug, touch hands or fondle genitalia, depending on how affectionate they feel. Chimps express intensity of emotion also when they are angry. They shake branches and glower, or when furious stand very straight and sway from foot to foot. Gestures communicate warnings as well. When a group leader spots a human in the vicinity, it will raise its arm and the followers disperse quickly leaving the leader to stare down the intruder.

Everyone agrees that animals can communicate how they feel (express their affective states), but there is disagreement about whether they can communicate anything else. Many experts suggest animals cannot communicate about objects unless they are immediately visible, about the past or future, or the location of food or other animals. In short, they consider them incapable of language—the highly organized system of communication which permits us to inform one another of a great deal more than emotional states. Other experts maintain that certain systems of animal communication seem to include features of the language system. There is the "all or nothing" view which excludes any creature below the human as a likely candidate to have a language, and the more evolutionary view which recognizes that features of language are used by some animals, and, incredibly enough, one of the animals is an insect.

Karl von Frisch, who won the Nobel prize for medicine in 1973, described how forager bees inform other bees of the location of food, a language-type message. The specific calls used by vervet monkeys to identify avian, mammalian, and snake predators seem to carry information as well as emotion. But many calls express an animal's emotional state only; they do not inform other members about conditions in the outside world—objective facts are not reported.

Detlev W. Ploog of the Max Planck Institute in Munich, a psychiatrist, identified twenty-six vocal sounds in the colony of about fifty squirrel monkeys he observed in his laboratory. Squirrel monkeys twitter constantly as they eat, and it was first thought they twittered excitedly because they were aroused by food, but it is now clear the monkeys are conveying a message of a desire for privacy. They like to maintain a distance between themselves and other squirrel monkeys during their meals, and regulate this distance by a constant twittering, whose intensity increases when one monkey feels another is too close. Purring, most often heard among the young monkeys, is also heard from grown animals, which purr as they smell and hug infants. Ploog calls these sounds "mood" indicators.

Faces can express feelings—frowns, tears, smiles are indications of anger, sorrow, and happiness. Generally, we can look at another's facial expression and guess the person's mood. But we don't know whether animals can interpret one another's facial expressions. Robert E. Miller and his associates studied this question in an experiment with rhesus monkeys. Two monkeys were restrained in chairs, unable to move except to press a nearby lever with their hands, or make head and facial movements. Miller showed two kinds of objects to the seated monkeys: desirable ones and frightening ones. But only one of the monkeys could see the objects. The other could see only the face of the first monkey. The second monkey had been trained to manipulate a lever which either brought the test object into the experimental room or removed it from sight. This "interpreter" monkey controlled the fate of the objects, deciding what to do by watching the observer monkey's face as it responded to the various objects.

Both the humans and the interpreter monkey easily recognized the fear expression. In the wild, submissive monkeys are furtive when they encounter a dominant animal. They take quick peeks of apprehension, glance away often, avoiding eye contact with the dominant animal if at all possible. The observer monkey produced just such expressions when it saw the frightening object. The interpreter, deciding those facial expressions meant bad news, promptly removed the undesired object with its lever.

A few years later the same study was repeated, but now with several isolate monkeys raised at Dr. Harry Harlow's laboratories in Madison, Wisconsin. The new results were even more instructive because of the method used to produce isolate rhesus monkeys. Shortly after birth, infant monkeys are separated from their mothers and placed in isolation chambers, where they are deprived of all visual and physical contact with other animals. After six months, the isolate monkeys suffer from a large number of deficiencies, particularly in their social behavior and locomotion. They spend the majority of their time clasping themselves. Placed with other animals, they neither groom, play, nor engage in any of the usual rhesus pastimes. Several attempts to reverse the damage failed. When isolates were placed with normal monkeys of their own age group, the normals attacked them, compounding the problems of the pathetic isolates further. The clasping, rocking and huddling behaviors increased.

On the day I visited the Wisconsin laboratories I watched the isolate monkeys stand erect, moving about on their hind legs instead of on all fours as is their normal position, and had a sudden, chilling impression these were tiny old men in cages. Dr. Stephen Suomi, a colleague of Harlow's, thought the isolates would be safe in the company of non-aggressive animals, but almost all the normal rhesus monkeys were hostile except for one group—three-month-old infants. They were too young to be aggressive; in fact, their only social behavior consisted of clinging to another animal. Suomi placed each isolate in a cage with a normal infant monkey, and soon the isolate huddled in the corner once again, but not alone now, for the infant hugged it tightly. Two weeks later, the isolate was hugging the infant. When the normal monkey started new play patterns, it encouraged the isolate to join in play. By the time the isolates were a year old, they resembled their rhesus friends and engaged in the same amount of exploration, locomotion, and play. When the isolates reached the age of two they were safely housed in a group cage and seemed fully recovered.

When Miller repeated his experiment, he included three isolate monkeys which he paired with three monkeys raised in the wild. Both the wild and laboratory raised monkeys can express and interpret facial expressions, and all groups can be trained to manipulate levers. The particular isolates used in this study had not undergone socialization with an infant monkey. They could not interpret, particularly, fear expressions, and were inclined to bring feared objects into the test space. In the reverse situation the isolates did not register normal facial fear; they could neither express fear nor interpret facial expressions of fear.

They lacked communicative skills because they had had no chance to acquire them. Isolate monkeys can only develop social behaviors by learning them with a group of other monkeys. Communication by means of facial expressions is one kind of social behavior which must be acquired. The fear expression on the face of the isolate was not like that on the faces of either the wild or laboratory raised monkeys. Since the isolate can be restored successfully to a range of normal social behaviors, it would be a particularly nice ending if the Miller studies could have included socialized isolate monkeys. One could say the restoration of the isolate was complete if this group produced and interpreted facial expressions accurately, something the untreated isolate failed to do.

"Tell me where the bananas are"

Emil Menzel, now at the State University of New York at Stonybrook, was interested not only in the facial expressions of chimps, but in what might be called "body language." When one of Menzel's unrestrained chimpanzees faced a snake, the fearful stimulus sent the chimp into a stew of emotion compared to the limited evasive glances of the restrained rhesus monkeys.

Above, a group of chimps at the Delta Regional Laboratories, Covington, Louisiana, 1968, locate and uncover a model of a snake in a test of communication.

Below, a monkey on Koshima Island adopts the innovation of washing potatoes.

Nadezhda Kohts compares the behavior of her chimp, Ioni, with that of her son, Rudi, ten years later.

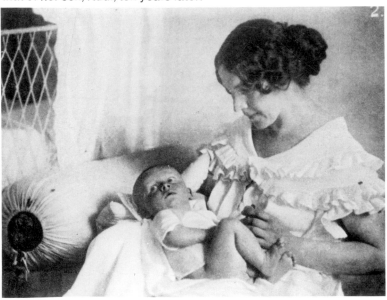

Keith Hayes with his chimp, Viki, who lights a cigarette and pours a cup of tea in a series of tests of imitation in the Hayes's home near Orange Park, Florida, 1940s.

Viki, at six, demonstrates self-recognition as she examines herself in the mirror.

Lucy, a home-raised chimp, communicates in sign language by making the sign for fruit when Roger Fouts shows her an orange at the University of Oklahoma, 1970s.

Infant chimp, Sarah, in her crib, 1964, at the University of Missouri.

Peony at Santa Barbara, 1973, prepares for her lesson by placing her name around her neck.

Two chimps at Santa Barbara, Sarah and Gussie, play with their first language device, a joystick, which produces a variety of sounds.

Elizabeth frolics with her trainer and by herself on the campus at Santa Barbara.

Above, in the classroom at Santa Barbara, Spring, 1975, a trainer works with Elizabeth who first reads the instructions, left, and then responds by "washing" the apple in a carton of water.

Above, trainer and Elizabeth both wear their names as Elizabeth writes a message to her teacher, fall, 1973.

But what kind of information can one chimp give another with body language? Can it inform about location? The forager bee can give such information to other bees. After a forager bee has made several trips to a rich food source, it performs a dance upon returning to the hive. On the vertical honeycomb, the forager starts to waggle. When its tail waggles upward, it means food lies in the direction of the sun; if it waggles downward, the food is opposite the sun. The faster the forager waggles, the closer to the hive is the food. The message can even be understood by the human observer, who, though wingless, can locate the food source as accurately as can the observer bees.

Monkeys and higher primates don't provide factual information to one another about the location of food or of a predator such as a leopard. They usually remain in the general area of food or predator while issuing "calls." But they cannot inform about distant locations as bees are able to do. Communication by visual gaze or circling is helpful to animals in the immediate vicinity, who can see what is happening; but authentic information about "where the bananas are" can be provided by a discoverer to others in the group even when far away from the location being described.

When Emil Menzel first began his studies of chimpanzees at Delta Regional Laboratories in Covington, Louisiana, in 1968, he wanted to know why chimps congregated —what made them form a group. He soon became convinced the chimps were communicating certain kinds of information to one another, a discovery which so intrigued him he lost interest in how groups were formed.

In the wild, chimps spend the major portion of their day with their close companions, traveling in search of food or to rest areas, but we know little of how all this activity is directed. So, using a large field laboratory and eight juvenile wild-born chimps, Menzel began his observations. The chimpanzees had lived as a group for a year, were all about five years of age and had a generally friendly feeling for one another. Three of the animals he had named Bandit, Shadow, and Belle formed a close unit, while Libi and Bido were a close pair, as befits their names. Polly, a loner, drifted freely between the two groups.

The chimpanzees were tested in a one-acre compound, where a variety of foods, toys, snakes, and other objects were hidden in clumps of leaves, behind trees, or in the tall grass. Usually, only one object was hidden at each test, and hiding places varied from day to day. No animal found any object by accident. Animals picked to be leaders were specifically those who could tolerate being separated from their companions. Not all could. Bandit and Belle were usually chosen, as they were not disturbed by the separation. After an object was hidden in the compound, the chimp leader was carried to a location, shown the hidden object, and returned to the group for a few minutes. Then the release cage was opened, the animals moved toward the cache with their leader, while the experimenters watched the chimps from an observation tower. To rule out the possibility that olfactory cues were helping the leader remember the location of the food, each experimental trial was balanced with a control trial. On control trials, though the leader left the group for the compound, it was not shown anything; the group searched aimlessly on these trials, eventually refusing to take part in them. But on the experimental trials, when the leader was shown hidden objects, the chimps were as eager to start as a pack of greyhounds.

Familiar group leaders were more popular than dominant ones, for when Rock, a large new male was introduced as leader, he could not attract followers. Belle was very successful in gathering a group around her, and Bandit was magnetic as well. Menzel pointed out the most successful strategy for attracting followers was for the leader to move surely and independently in a consistent direction, a tactic seldom, however, adopted. Perhaps a leader could behave with such certainty only if it had a clear and positive evaluation of the food. Leaders, if given the opportunity, almost always refused to go after the food by themselves. Bandit usually had a tantrum when left alone after being shown hidden food. He never returned to the cache by himself, but instead screamed, rolled on the ground, and tore at his hair, finally running off and clinging to a tree for security. Solitary leaders remained sulky or furious, but when the group cage finally opened and the other chimps emerged, the leaders speedily recovered their aplomb, ran to embrace one of the followers and were off like darts in the direction of the food.

When a leader's hair bristled, and he issued alarm calls and hoots on returning to the group, something fearful was anticipated by the waiting chimps. When snakes had been hidden, all the animals were pilo-erect, their hairs bristled, as they emerged from the release cage. The animals were in a state of high arousal, and were intrigued and excited about going to the fearful spot. On a number of occasions, the snake was removed from its hiding place shortly after being shown, yet the animals returned to the correct spot, anxious, sometimes throwing sticks in the vicinity.

Menzel believes locomotor postures and visual (eye movements) pointing give enough information for an animal to find the location of an object. The speed of travel toward the location, and the changes in acceleration as the cache is approached, cue followers in evaluating the hidden object. Because Menzel's chimpanzees did not at any time go to a food or a snake cache without their leader, one is inclined to think information about a location was not given while the animals were in the re-straining cage together, waiting to be released. The chimps were so extremely dependent socially they could not move without a leader, nor would the leader move without followers. And when Belle or Bandit sometimes preferred a siesta to leadership, the others poked and cajoled until the leader returned to duty. Did chimps give information about location giving factual information? Or is the leader a compass without which no other chimp could find its way? In one study of dual leadership, Menzel discovered that the group followed Belle if she could lo-cate the larger amount of food, or Bandit, if *he* could. Chimps followed the leader to a generous hideout first, before meandering over to a lean one. Now it was *not* the leader, but the amount of food which took priority. How the chimpanzees communicated this information remains a puzzle.

Hans Kummer, a primate ethologist and author of *Primate Societies,* suggests that without symbolic communi-cation animals cannot indicate "where the bananas are," they must *lead* animals to the bananas. Telling where ob-jects are located requires propositional communication, that is, a language statement. Telling does not require a *spoken* language, of course, for the bee tells location through its waggle dance.

If young children took part in the Menzel studies, the children could be induced to locate a hidden object by simply following the leader, quite as the chimps did. Or the leader could decide to rest, and instead of leading, draw a map of the area with the location of the hidden object clearly marked so the others could find the cache on their own. Or the leader could give language information: "The Milky Ways are under the palm tree, four giant Cokes behind the center rock, and watch out for the snake hidden in the weeds." Children can show or tell in giving information about where things are, but the chimps seem only to show.

But chimps aren't incapable of using symbolic communication when informing others about the location of hidden items. If taught names for the several hiding places, "red rock," "small house," "brown tree," "brown rock," "big tree," and so on, as well as the names of objects hidden, "snake," "bananas," "apples," "Cokes," then a leader, after being shown the cache could write the information on a bulletin board. He could inform them of the "apple under red rock," "banana behind brown tree," "female in small house," then retire for the day. Other kinds of general-interest information could be relayed as well—news of interest to other chimps in the compound such as the location of comfortable nesting trees, names of females in estrus, or what the experimenters were up to. These types of experiments will be done at a new chimpanzee facility near the University of Pennsylvania. Emil Menzel and David Premack are collaborating on this important group of studies.

Symbols: used by humans and (perhaps) beasts

There is a difference between communication and language which can best be clarified by examples. My dog, Bistro, can understand voice tones and general excitement to mean she is about to go for a walk. She also obeys a few simple commands, sometimes. If I tell Bistro we're going for a walk, she may not understand the word "walk," but aroused by the tone of voice or the sight and jingle of her leash, she runs to the front door in preparation for the daily jaunt. Am I justified in assuming she actually understands the word "walk"?

Winthrop Kellogg reported that at eighteen and a half months his son, Donald, had a comprehension vocabulary of 107 words while Gua, the chimp, at sixteen months had 95 words in her comprehension vocabulary. Some of the phrases and sentences were: "no-no," "kiss-kiss," "come here," "shake hands," "sit down," "open your mouth," "get up on the bed," "lie down," "show me your nose," "blow the horn [car]." Probably both Gua and Donald were responding holophrastically to the phrases, a key word providing the cue for understanding. To claim that an animal understands all words in a sentence, we must show it is able to follow directions when each word in the sentence is contrasted with every other in a series of sentences. If an animal is told to "lie on the bed," and does so, the response can mean very little. Perhaps it habitually lies down on a bed, or it associates the word "bed" with lying down. To illustrate true semantic comprehension, the animal has to demonstrate it can lie on the bed, sit on the bed, lie on the chair, sit on the chair, lie under the bed, lie under the chair, sit under the bed, sit under the chair. When the animal or child can make all the distinctions required in the series of sentences by lying or sitting, on or under, bed or chair, then we can claim semantic comprehension.

About forty years ago, John Wolfe, an experimenter at Yerkes Laboratories, taught a small group of chimps different meanings for a few colored poker chips. When a chimp wanted to engage in a particular activity—to eat, drink, play, and so on—it placed a colored poker chip in a kind of slot machine, and received what the particular poker chip represented, food, water, games, etc. Bula

"To illustrate true semantic . . ."

and Alpha were included among the six chimps in these studies—the same Bula who was housed with the same young chimp Alpha, whose physical development the Jacobsens studied in 1932. All the chimps were young—Bula and Alpha were just two years old. (In the course of reading, I ran across a study where Bula and Alpha were subjects once again. But now they were thirty years old! For several minutes I was lost in fantasies about what had happened in the intervening twenty-eight years.)

In the first study, when the chimpanzee placed a blue chip in the modified slot machine it was permitted to leave the work area and return to its home cage; but when the yellow chip was inserted Wolfe would play with the chimp for a short time, then return the chimp to its cage. Wolfe tried to encourage Bula to use the blue chip by a clever ploy. He placed her cagemate within earshot of the test room, but out of Bula's sight. When she heard her friend vocalize she was desperate to join him, looked about until she finally spied the blue chip, and inserted it in the slot. She climbed onto the experimenter forthwith, ready to be carried out of the room. At another time, Bula used the chip in a completely unplanned situation. Annoyed by a photographer in the test room, she wanted to escape the flash of cameras, which frightened her, and she promptly offered a blue chip for her release. Once, when Bula was busy inserting white chips in a machine which dispensed grapes, Wolfe surreptitiously placed a rat on the table. When Bula spied it, she swiftly located a blue chip and obtained her freedom.

Bula requested play, a game of tag with the experimenter, by using a yellow chip. On one occasion, while she was again busily collecting grapes with white poker chips, Wolfe quietly placed a variety of chips on the floor. Bula selected a yellow chip, inserted it in the slot and came running toward Wolfe. Seeing her hurtle toward him, he tried to avoid the collision, whereupon she promptly tackled him. For Bimba, a chimpanzee with his own ideas about recreation, play consisted of being swung around in a circle in exchange for a yellow chip.

All the chimps in the study learned to select the colored poker chip which dispensed two grapes rather than the chip which dispensed only a single one. Two chimps which had been deprived of food or water for a considerable number of hours chose food chips if they were deprived of food, or water chips if they'd been without water. Although all animals could distinguish the different grape chips, a few animals had trouble learning the meaning of the activity chips. As with Köhler's chimps, there was a variation in learning ability.

The poker chips had no resemblance whatever to the objects and actions they "named." The relation between the chips and their referents was entirely arbitrary, one of the traditional requirements for a symbol, but Wolfe hesitated to call his colored chips symbols. It is difficult to understand his caution today, for it seems Wolfe taught his chimps something far more advanced than simple association.

Perhaps the ability to symbolize is not so special; man seems not the only symbol user. How else are we to account for chimps that learn sign language, as Washoe and other chimps have done? Or chimps that write sentences as Sarah has? Washoe and Sarah are chimpanzees that were each taught about 130 symbols which they used in lessons and short dialogues with their human trainers. We may be defining some concepts too narrowly, for symbols are probably used by other creatures besides us, though we use an enormous vocabulary of symbols in complex structures called sentences.

5

washoe
learns
signs

No one has tried to teach chimpanzees a spoken language since Viki failed to speak. In spite of the human surroundings, the daily exposure to speech, the careful daily lessons using the most sophisticated procedures, intermingled with love and friendship from Keith and Cathy Hayes, Viki could not learn human speech. The failure was totally convincing to scientists, dissuading anyone from saying, "If only they had been more strict . . . or more permissive. . . ." No one attempted another language project with chimpanzees for almost twenty years. Then some psychologists had a simple but overwhelming insight. They saw that language must be distinguished from speech. In common usage, language has been equated with speech, but any formal system of communication is a language. If a chimp could not learn speech, it might learn another form of language. Language can be expressed in Sign Language, writing systems, or any other system of arbitrary symbols. Talking is *not* the only way to conduct a conversation.

Had Cathy known what we now know, she would have been cheered by Viki's response to pictures. For Viki used pictures, spontaneously, to request and obtain the real objects they represented. For instance, at the age of four,

"Language can be expressed in Sign Language . . ."

Viki pointed to pictures of drink advertisements, said "cup" and led either Keith or Cathy to the kitchen for a drink. She pointed to a candy-bar advertisement in a magazine, produced a few "food" barks and set off for the pantry to find some real candy. Scientists call this use of representations "icons." An icon resembles the real object it represents and differs from a symbol in that a symbol bears a completely arbitrary relation to the real object. The written word "cup," the spoken word "dog," have no resemblance to a real cup or dog. Viki was unable to learn spoken words, was not taught a written system, but invented her own system of obtaining objects she wanted.

She used a small number of pictures, "icons," to communicate her desires. The icons were readily recognized and understood by her human parents.

Had Viki been taught a system for communicating with pictures, she could have exchanged information with the Hayeses. Imagine the likelihood, after a few weeks, that the pictures have become faded, crumpled, torn, so "cup" is reduced to a handle, "bottle" a bare silhouette, and "outdoors" wadded into a ball. Yet Viki still uses the pictures correctly. Now they are no longer pictures; they are symbols. The relation between the name and the thing has become arbitrary; the resemblance between name and object is nonexistent. Viki used pictures as icons for objects she wanted and her requests were understood because icons resemble real things and chimps and people can agree on their meaning. But a symbol bears no resemblance to its referent, as a wadded paper ball has no likeness to "outdoors"; it demands more from memory and becomes the ultimate label.

Because they were impressed by the large number of gestures chimps use in the wild, Beatrice Gardner and her husband, Allen, at the University of Nevada in Reno, decided to teach their chimp Washoe to communicate with gestures. They considered, at first, inventing a language of gestures, but discarded the idea as too demanding and unnecessary since the well-established language system of signs, long used by the deaf, was already available.

For many years American Sign Language was regarded as something less than language, for it was seen as a language of icons, in which signs resembled the objects they named. Though many signs are iconic, most signs are symbolic, as are most words in spoken language. Sign is neither a derived form of English nor a novel language but a very old language system with a long history, and a vocabulary which presently consists of about two thousand signs. When they converse, signers vary their hand shapes, the location of their hands, and the direction hands move. Signs can be made by one hand or both; sometimes each hand designates a different word. Each sign usually represents a word, though signs can mean several words, or perhaps a complicated idea. Though Sign Language has neither suffixes nor prefixes, neither articles nor the verb "to be," it manages nicely. Signers can communicate some ideas very swiftly, using the simple technique of varying the speed of a repetition. When a signer repeats the "talk" sign briskly, he means he "talked" to a lot of people, but if he "talked" to a single person for a long period of time, he repeats the sign slowly. To indicate future time, a signer points forward slightly with his index finger, or reverses the motion to mean past time. In describing a scene, a signer uses the space in front as a kind of stage for locating the characters in her drama. People and objects are assigned places on the imaginary stage and when the story finally begins, the "listener" knows the actor by simply looking where the "speaker" points on the imaginary stage, constructed by the speaker for the listener's convenience.

Washoe, in the tradition of home-raised chimpanzees, was kept in a house trailer. She had all the amenities of home there, a legitimate bathroom, a fully furnished kitchen, and her own bedroom. The trailer was located within a fenced yard containing trees, shrubs and flowers, and outdoor play equipment. To enrich her home life, she was taken on frequent outings to other homes, ponds, and so on. The Gardners wanted to provide Washoe an interesting world to communicate about, but they did not intend to simulate family life. From the time she was bedded down for the night until she awoke in the morning, she was alone. But during the day she had the company of at least one person.

All Washoe's teachers taught themselves some signs and tried to create a Sign environment for Washoe. They signed to one another in the chimp's presence, hoping she would imitate their gestures, as a normal child is expected to acquire spoken language, by observing and imitating. At first she was rewarded with things she was partial to when she made any gestures at all, but later she was "treated" only if the gestures were similar to those of Sign. The Gardners encouraged their chimp to imitate by tickling, smiling, and feeding her preferred foods whenever she mimicked their signs, but this method for building a vocabulary was slow. To speed things up, the Gardners began to move Washoe's arms through the desired gestures, to guide her through the motions. Fortunately, this assistance stimulated an increase in the sign vocabulary. If Washoe repeated a guided sign independently, she received the object that the sign named.

Washoe's sign for "toothbrush" seemed to come through delayed imitation. She had to brush her teeth after meals, a routine she at first despised, but after a few months she acquiesced and even enjoyed it. She was not allowed to leave her high chair unless she first brushed her teeth. The Gardners signed, interminably: "First, toothbrushing, then you can go," but results from Washoe were meager. Several months later, on visiting the Gardner home, Washoe wandered into the bathroom. She observed the dental equipment, especially the mug of toothbrushes, and signed "toothbrush." It was the first time the Gardners had seen their chimp produce the sign. Her imitation of the sign for flower was also delayed, for though she was shown the sign in the presence of both real and pictured flowers, she did not make the sign herself until she later took a walk in the garden.

It was in the context of tickling that Washoe learned "more," for when she made the sign the Gardners resumed tickling her. Soon she used "more" to request the continuation of other play activities and finally signed "more" when she wanted more food. "Open" allowed the chimp into previously forbidden places, and later she signed "open" to refrigerator doors, cupboards, jars, until she finally requested the water faucet be "open" as well.

In the tenth month of the project, when Washoe possessed a vocabulary of no more than a dozen words, she did something totally unexpected. She spontaneously signed "gimme sweet" and "come open," combining her words innovatively. The Gardners had often signed a series in the chimp's presence, but had never deliberately taught her to combine signs. The procedure was adopted by Washoe herself, and it later developed into combinations of three to five signs, strung together.

So many of Washoe's signs were used in an appropriate context that it was easy to enter into signing with her. She signed, "Roger you tickle" and "you Greg peekaboo," or she wished to "go in," "go out," "in down bed." She was fond of signing "sorry" and frequently gestured "please sorry," "sorry dirty," "sorry hurt," "please sorry good," and "come hug sorry sorry." "Sweet drink," bottled pop, was popular as well, and she requested some with "please sweet drink," "more drink please," "sweet drink more," and so on. Her most frequent signs were "please out," "come open," and "more tickle," which she used in short dialogues with her human companions.

Though the chimp reached the stage of signing three or

more signs in a sequence, as in "please tickle more," "come Roger tickle," "you me in," "you me out," and others, Washoe continued to sign them in reverse or scrambled order. At the age of about four, she had a vocabulary of 85 signs, some of which she used in individual signs and others she joined to form three or four in a sequence. She and her trainers engaged in short, simple dialogues.

The Gardners considered Washoe's environment rich in comparison to those of most laboratory chimps, but impoverished compared to that of a middle-class child. Washoe did not have fluent signers as teachers, or peer signers to play with, detriments they have now corrected in replicating the Washoe study with two infant chimps being taught signs by knowledgeable signers who are deaf, and for whom Sign is a first language. Armed with a vocabulary of 130 words, Washoe moved to Oklahoma to live out her adolescence and the rest of her life with a large group of chimpanzees.

Will chimps have dialogues?
with each other?

In October of 1970, Washoe settled down with a number of other caged chimps on a kind of farm in Norman, Oklahoma, established by Dr. William Lemmon, a clinical psychologist who has a fairly large collection of primates. I visited Oklahoma in 1973, on a bitingly cold February day, and accompanied Dr. Roger Fouts on his teaching rounds. Fouts had worked with Washoe some years earlier while a graduate student, and now teaches signs to several chimps in the hope they would have dialogues with other chimps. The first stop on our rounds was at Lucy's. One of the chimpanzees at the farm, Lucy, was adopted by Lemmon's secretary and raised in her home. As we entered the driveway I heard a chimp call. It was Lucy welcoming us from her private apartment, a large cage built on the roof of a house. Lucy spends a great deal of time with her human family in the main portion of the house, eats politely with knife and fork, and examines the faces of all guests. She refused to sign with Roger while I was in the living room, preferring instead to run her enormous index finger around my mouth and nose as I chatted with her teacher. When Roger handed Lucy a dish of orange segments, she fed me, but carefully, like a nurse, making sure I'd swallowed before pushing another segment into my mouth.

A day later, Lucy was less interested in me and began to demonstrate her signs. Roger brought several objects in a medical bag, presenting each of them to the chimp with the question "What's this?" It was a lesson he repeated with the same enthusiasm to every chimp taught signs. Lucy signed "fruit" to an apple, "comb" to a comb, "dog" to a stuffed animal, "pipe" to a smoking pipe, "baby" to a doll, "flower" to a plastic flower and so on. The majority of her sign sequences were requests for tickling, "tickle Lucy," "me tickle," and "tickle Roger tickle."

Later we visited a young chimp named Allie being raised by a social worker. The typical suburban tract house we drove to looked ordinary enough from the outside; and so did the interior, until we entered the guest bedroom. Here, sitting in an enormous and beautiful cage as permanent guest, was two-and-a-half-year-old Allie, watching his color TV set with some attention. Roger once again opened his object bag and went through the drill with Allie as we all sat comfortably in the living room. Allie demonstrated his sign lessons with alacrity. Roger mentioned that Allie knew about ninety signs, whereupon

Allie signed the sequences "you tickle me," "tickle Allie Roger," and "Roger tickle Allie." Then Allie produced the sign for "dirty" associated with going to the bathroom. Roger said that Allie frequently made the sign but seldom performed, but this time he did, and from the torrent I heard from the distance, they were just in time. As Roger and I left Allie in his cage, a young kitten entered through a small flap to keep Allie company while the foster mother worked. We left the TV set on.

We drove to the farm, an assortment of old buildings and barns situated on a slight hill. On the far side of the entrance was an artificial lake, where three islands sat like dumplings in a muddy stew. In the summers, the chimps are moved to one of the islands, but they were now in their winter quarters. We walked into the main caging area and faced about fifty chimps of all kinds—spoiled, one-time circus performers; skid-row types; home-raised chimps too large to remain home—now they all had a permanent home. Roger leashed Washoe and we walked to a comfortable barn, where Roger and Washoe sat on the rug for sign lessons.

Larger animals were usually taken singly from the caging area to a barn for lessons. Washoe was not learning new signs but was being reviewed on her old vocabulary to be sure she did not forget. When Roger showed Washoe her favorite fruit, she became very excited and began a long series of aroused and quick signs which Roger translated for me as Washoe gestured wildly: "More please fruit, more gimme, more you, please more gimme, please fruit, more you gimme fruit, gimme more, gimme fruit gimme, Roger." She halted occasionally and eventually slowed to a complete halt. The signs, though they occurred in a disorganized fashion, under great emotion, were meaningful in the situation. What is curious about the chimp's use of signs is the constant repetition, particularly when the chimp is excited. Washoe signs, "More more more more sweet drink," "Out, out out please out," and she and others will produce the sign "dirty" several times, or "hug hug hug hug hug," "red red red," and so on. Whether children acquiring Sign Language do the same thing I don't know. This type of repetition, however, is a common feature of Sign, though not in the degree found in chimpanzees.

Does this repetition of signs tell us anything about the structure of the chimp's mind or is the process an emotional one? There does not seem to be much evidence of this kind of repetition in children learning speech. And there are other differences between the way chimps sequence signs and children sequence words. As soon as a child begins to use two-word sentences, at the age of about two or so, it uses a special pattern in producing them. The child will say, "Allgone milk," "Allgone daddy," "Bye bye sock," and "Bye bye mommy," but would not say the reverse, "Milk allgone" or "Sock byebye," or tack on an extra word "allgone milk allgone, sock byebye byebye." The child does not imitate adult sentences, producing as scientists used to think, telegram sentences. The child has a pattern for producing sentences even as early as two.

When Washoe reached the two-sign level her signs (as well as those of the several other chimps learning Sign), were ordered differently. The early sentences vary from "Please gimme" to "Gimme please" or even "Gimme please gimme." There are reversals and repetitions rarely found in the early constructions of children. If there is a pattern at the two-sign level, it hasn't yet been found. Without some kind of order, it is not possible to understand word or sign sequences. "Tickle Roger Washoe" has meaning only if we know a great deal about the traditional tickling situation, that is, Roger always tickles Washoe. But, if Washoe wanted to tickle Roger for a change, she would have no sign sequence to help her out. In languages, word order is essential in both the production and understanding of messages; in Sign Language, order is equally important.

Because of the considerable current interest in the language work with chimps, newspapers and magazines are burgeoning with articles, which, though entertaining, are sometimes inaccurate. Some inaccuracies come about because writers draw incorrect conclusions about chimp language performance. One such misunderstanding concerned the reported signing of "water bird." One of the chimpanzees being taught signs knew both "water" and "bird," but had never before seen a duck floating on a pond. Upon seeing a floating duck for the first time, it signed "water bird." The writer concluded the chimp meant "water-bird" and praised the animal for inventing a clever new word. But it is the writer who assumes the words are hyphenated, not the chimp. Suppose the animal had signed "bird water," reversing the signs as chimps are prone to do? Are we as apt to assume these are hyphenated, are intended to represent a single word?

Perhaps all the chimp meant was "water" "bird," two separate labels for things it recognized; a hyphenated, new word probably had no more meaning to the chimp than to the child who was once asked why the hyphen was placed in "bird-cage." "Why, that's for the bird to perch on," she replied.

6

teaching sarah
to read

In 1797, a young boy was discovered by hunters in the Caune Woods of France, his body completely covered with scars. He smelled anything handed to him and, instead of speaking, produced only guttural sounds. At first, he was thought to be "wild," but doctors who examined him suggested he was extremely retarded. The boy was brought to a young physician at the Paris Institute for deaf mutes, Dr. Jean-Marc Goddard Itard, who was soon both smitten and challenged by the boy's savage behavior. Convinced he could educate the boy, Itard assumed his complete care. He named the boy Victor, and introduced him to the comforts of civilization, a bed, clothing, cleanliness. But Victor adjusted slowly to the dramatic changes in his way of living; at first he preferred to scrounge outdoors for acorns, potatoes, and raw chestnuts rather than eat cooked food. Later, he came to enjoy his new comforts. When Victor felt more at home, Itard tried to teach him to talk. But the effort failed.

Suspecting Victor's vision was more acute than his hearing, Itard next tried an original approach to language instruction that did not involve spoken words or sounds. He began by printing on two blackboards two lists of the same words, only not in the same order. When Itard pointed to a word on his board, the boy was to point to the matching word on his own board. When Victor made a mistake, Itard would point to the first letter of the word, while Victor did the same, and proceed to match letter to letter until they reached those letters that did not match. And in this simple fashion Victor learned words such as "book," "key," "pen," and "box."

One day Itard locked a book in a closet, then gave the word "book" to Victor, who immediately went to fetch the one he was familiar with. He looked everywhere, and, unable to find the particular book he recognized, was frustrated. Though there were other books in the room, Victor did not pick any of them, because he had not yet managed to transfer his word "book" to other books. Victor lacked the ability to transfer spontaneously, something Washoe could do after she learned the sign "open." In slow steps Victor was taught to apply the word "book" not only to the original, but to many books, and in time he applied the word to all books.

In later vocabulary lessons, Itard constructed metal letters of the alphabet, spelled the word "milk," then poured himself a glass of milk. Scrambling the metal letters used to spell the word, he invited Victor to follow his example. When the boy spelled "milk" correctly he also received an icy glass of milk, of which he was very fond. Shortly after the lesson, Itard took Victor, as was usual, to visit a friend who liked the boy, and frequently offered him milk and other treats. Before leaving home, Itard noticed that Victor rummaged about in the container of metal letters but he thought nothing of it until, upon arriving at the friend's, Victor triumphantly removed some letters from his pocket and spelled the word "lait." It was the first time

Victor had shown any real understanding that a word stood for an object.

Victor seemed entirely without language ability when found, either because he was an isolate, unsocialized child, or because he was deficient and for that reason had not acquired language. He symbolizes the many thousands of deficient children who, though raised in normal homes rather than living alone in the wild, do not spontaneously learn to speak. Some of them are severely retarded, others are autistic and cannot respond to the social life which is essential for language learning, and still others are blind or deaf. Many of the children cannot utter sounds, or utter incomprehensible sounds. They need to be taught a language based on some other system than speech. Quite by accident, in developing a language for a chimpanzee named Sarah, my husband, David, presently at the University of Pennsylvania, introduced a system which can be taught to profoundly deficient children. Visual and manipulable, the language has been and is being used to teach children like Victor words, their meanings, and simple sentences.

When the language system was initiated at the University of California at Santa Barbara in 1967, I volunteered to help teach Sarah. Each symbol in Sarah's language system is of a unique shape and color, constructed of plastic, and is backed with a piece of metal that adheres to a magnetized board. Sarah wrote by placing the plastic symbols on the board in vertical sequences, as was her preference. Her vocabulary was carefully chosen by David as "exemplars" of language he hoped Sarah could learn—nouns, verbs, adjectives, pronouns, and quantifiers. Besides, she needed to learn several kinds of sentences—questions and imperatives, compound and complex. Finally Sarah was taught to write sentences in an orderly sequence. This language study differed from others because David had analyzed some of the basic requirements for human language first, and then attempted to teach them to a chimpanzee.

Ten years earlier, when the language project began and the laboratory was located at the University of Missouri, language was produced by a device based on sounds—an instrument called a "joy-stick," which produced "speech." It looked like a globe with a movable lever, and changes in sounds, based on organ tones, could be made by moving the lever by hand in different directions. Sounds could vary in pitch, length, and so on if the grip or position of the lever was changed, just as vocal sounds vary when the tongue, teeth, and lips work to change the quality of sounds in words. After the "joy-stick" was built, it was given to the chimps and a record was kept of the movements they preferred to make—a kind of babbling done by hand. But the mechanism was troublesome. David soon abandoned the idea that language needed to be based on phonology, on a system of sounds; he was convinced this model of language presented a needless complexity for chimps and tried a new system, based not on sounds but on colored visual shapes. The new language was taught to the chimps in 1967, after the experimental program had moved to the Santa Barbara campus of the University of California. There are probably not many campuses where young chimps frolic in the sun with college freshmen, and the two African-born chimps involved in the project, Sarah and Gussie, lost their quiet apartment in Missouri, but gained a more sociable way of life, and a new language.

Most of the early work with the chimps centered around

pilot studies using various interesting fruits, so that Sarah, for instance, could both solve her problem and *eat* it. In moving to a language of colored plastic shapes as names for a host of non-edible objects, Sarah needed to be taught what to do with objects she could *not* eat. For she had an unpleasant habit of taking glasses and pens out of pockets, and rushing to the back of her cage, where she methodically took them to pieces. She never returned anything, and it would have been impossible to exchange words and objects under the circumstances. It was necessary to reintroduce her to the art of barter.

Two graduate students were conducting research, making it an ideal time for me to acquaint myself with Sarah, but she resented my presence in her classroom. She looked at me and immediately started to hoot and display fantastic chest thumpings. The noise was deafening, intended as it was for the Gombe forests, not a small university campus. It reminded me of an initial experience with chimpanzees in 1954, when we went to Yerkes Laboratories for David's first research position. We spent the first night in the guest apartment on the grounds and awoke in the night to the sound of shrieking. The shrieks increased in volume, surrounded us, echoing and reechoing until even our intestines were vibrating. Sixty animals shook the bars of their cages in unison. After several minutes, although the thunderous noise subsided, my terror remained.

With Sarah, I sat courageously on a tall stool while she next reached through the wide rungs of the cage to grasp the stool leg, trying deliberately to unseat me. She made several attempts and each time I returned the seat to its original position, making it clear I intended to stay. After a few days she relented and soon grew friendly. She learned to take and return toys and to exchange one toy for another. At the end of barter lessons, she would eat a cup of fruit cocktail, which she loved, and we would kiss one another good-bye. Sarah liked to kiss, placing her stomach securely against the bars of the cage while protruding her mobile lips.

As an occupational therapist, I had worked with handicapped and retarded children. Sarah was not unlike a child to me. David and I would discuss what Sarah should learn, then decide on the simplest way to teach the lesson. I would try the procedure for a few days, pilot the technique, then decide whether the lesson was too difficult, or whether it should be added to the classroom repertoire. When the language project started, Sarah sometimes played hooky by refusing to begin her lesson, ended a lesson early by stealing words, food, or both, or threatening her trainers. Since some of the trainers, notably Mary Morgan, worked in the cage with Sarah, she was never deprived of food. It would have been unwise to deprive her, then hope to have her remain friendly. When Sarah became sexually mature and her physical strength increased, the trainers taught through a large porthole in the cage, on each side of which was a wooden shelf to hold the "words." Directly above Sarah's shelf stood the magnetic writing board.

A key is a key, not a clothespin

An animal can notice that two bananas are the same, that a banana and an apple are different, without knowing the names of the objects it is comparing. In the early tests of match-to-sample, for instance, no one taught the chimp Ioni the names either of the colors or of the complicated shapes he matched. He did, in fact, match subtle color shades without needing to say, "This lime green is the same as that." Sarah could also match objects without learning names.

Faced with two cups and a spoon, she learned to place the two cups together, and later, when she was given two corks and a key, two pencils and a pen, one spoon and two clothespins, and so on, she placed the matching objects together. When, in theory, she could have gone out in the forest and placed like objects together indefinitely, she was taught to place the plastic word "same" between the spoons and "different" between spoon and cup. Now Sarah was tested on a host of entirely new objects, which she also labeled "same" or "different."

This simple sequence can be translated into a question. In its simplest form, any situation which contains a missing element can be a question: Ioni without a pen to write with, Rudy without a hammer to pound his nail, two cups with an empty space between them. These are all situations with something missing, with a blank space, ready to be filled. The missing element (the blank space) was replaced by an interrogative sign, and Sarah was asked:

What is the relationship between a key and a key?
What is the relationship between a key and a cork?

Sarah's words "same" and "different" were provided with both sentences, and Sarah learned to remove the "?" and replace it with the correct word:

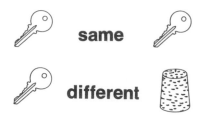

We could also ask:

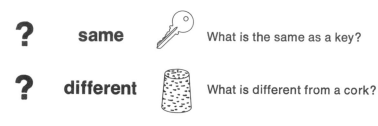

Once again Sarah removed the "?," replacing it with cork or key in the appropriate space.

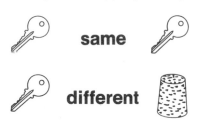

Before Sarah could answer "yes" and "no" to questions, the trainers taught her the meanings of these words she had not yet learned. And she learned the words in the context of a small tea party at which she and Mary Morgan sat at either end of a table. A fine spread, consisting of six varieties of finger foods—small rounds of bread and crackers, each topped with honey, jam, or peanut butter, were offered. The trainer invited:

```
┌─────────────┐          ┌─────────────┐
│ Sarah       │   or     │ Sarah       │
│ take        │          │ take        │
│ honey       │          │ jam         │
│ cracker     │          │ bread       │
└─────────────┘          └─────────────┘
```

followed by two messages at the same time:

```
┌──────────────────────────────┐
│ Sarah            Sarah        │
│ take             take         │
│ peanut butter    honey        │
│ bread            bread        │
└──────────────────────────────┘
```

Sarah took what she was thus offered and ate.

There is an early form of negation in almost every language when the child places a negative particle in front of the simple sentences he wants to negate, as "no daddy spank," "no go car," "no go bye-bye," and so on. Sarah learned the negative in this fashion. Mary changed her invitation to:

```
┌──────────────────────┐
│ Sarah      No        │
│ take       Sarah     │
│ honey      take      │
│ bread      jam       │
│            cracker   │
└──────────────────────┘
```

When Sarah reached for some jam cracker, Mary prevented her from taking it. Sarah soon learned not to take

foods in negated sentences. But, when there were no negatives, Sarah helped herself to both foods. Having learned
"no" she was taught "yes" in the context of another form
of the question, one that required a "yes" or "no" reply.

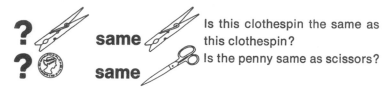

She placed "no" in the appropriate space, "yes" before
like items:

When being taught questions, Sarah often took the
day's stock of words, placed them on the floor, wrote out
several questions carefully, and just as carefully answered them. Later, on every third or fourth lesson, she
requested or simply took a large number of words, then
asked and answered several questions by herself. Sometimes her answers were incorrect. Of the eight or nine
questions she posed herself, at least two were answered
improperly. Though Sarah found the negative an aversive
word, she even asked: "What is an apple not?" and replied, "Bread."

In certain lessons, writing was quite innovative. Though
most of Sarah's lessons in reading and writing occurred
on her magnetic board, there were many occasions when

"When Sarah reached for some . . ."

it was more convenient to write on her work table. The sentences consisted of a mixture of words and objects:

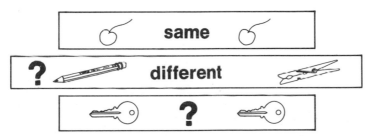

These types of sentences, called hybrids, are also used by children when they want things they cannot name. The children intersperse their speech with pointing, or use physical gestures to indicate their wants. Travelers in foreign countries, with limited knowledge of another language, can manage by interspersing actions and words, using adult hybrid sentences that combine icons and words. The hybrids worked equally well with Sarah.

Brown is the color of chocolate

How do we teach a chimp that, in looking at a three-dimensional red ball, the color red is to be its salient feature, and its round shape and soft texture should be ignored? One technique was to give Sarah a variety of objects different in every way except one: they were all the same color. The ball was red, the toy car was red, and so were the candy Lifesavers. To teach the color yellow, she was given a yellow toy block, a yellow crayon, a yellow carnation, Cheezits, and so on. The trainer placed a red ball along with the words "give," "Mary," "Sarah," and "red." When she wrote the correct sentence in the correct order on the board:

```
Mary
give
red
Sarah
```

Sarah was given the red ball. Next, the same thing was done with a yellow object and the word "yellow." Sarah alternated yellow and red trials until she had experienced six red and six yellow objects. Sometimes the sessions lasted quite a long time as Sarah kept and played with some of the objects, but she always returned the items when asked, except for the yellow Cheezits and red Lifesavers. Those she did not return. We had a stock of replacements.

In transfer tests, Sarah named correctly an entirely new set of yellow and red objects she'd never seen before. She even identified little cards painted red and yellow. Exactly the same procedure taught Sarah "round" and "square," instances of shape, and two instances of size, "small" and "big."

When she identified "big" and "small," "round" and "square," she was ready to be taught "color of," "shape of," and "size of." She learned to complete the following:

Red   ?   Apple
Yellow   ?   Banana

She was given the word "color of," which replaced the interrogative markers. In order to complete:

Red   ?   Banana
Yellow   ?   Apple

she was given the word "not color of," constructed by gluing the negative she'd already learned in front of the word "color of." And on her transfer tests Sarah did very well, replying to such questions as

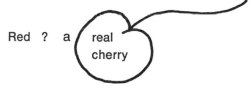

Red ? a (real
            cherry

quite correctly. Since many of the objects were not named in these tests, most of the sentences were written out on Sarah's desk, in hybrid fashion. It is certainly a measure of Sarah's flexibility that she moved from board to table, from words alone to a mixture of words and objects, without confusion. After learning the class concepts of size and shape, Sarah could answer:

Red   ?   Apple
Round   ?   Ball
Small   ?   Marble

correctly inserting her alternatives, "color of," "shape of," and "size of" in the appropriate sentences.

The advantage of teaching Sarah the concept of a color class arose when Sarah was taught new color names. Now, to teach "brown" and "green," we no longer needed a bagful of green and brown objects. Sarah was simply told in writing:

> Brown
> color of
> chocolate

Did she really understand what was meant by "color of"? The trainer showed Sarah four wooden discs of various colors, only one of which was brown, and wrote:

> Sarah
> take
> brown

And Sarah took the brown disc. When the trainer wrote:

```
Sarah
insert
brown
red
dish
```

she placed one colored disc (of four) into a colored dish (of two), placing the brown disc into the red rather than the green dish.

One very important phenomenon occurred when instructing Sarah on the color brown. She learned it in the absence of a brown object. When the sentence:

```
Brown
color of
chocolate
```

appeared on the board, there was no actual chocolate to be seen; only words appeared on the writing board. Later, when told to "take brown" she took an actual brown object. How was Sarah able to do this? The ability to think of things not immediately present is called displacement and is considered a capacity unique to man. But Sarah was apparently able to think of chocolate (she was fond of it), even though chocolate was not immediately present. To do this, she must have been able to form a mental image of some chocolate, compare the color in her image with that of the brown disc, and match the two colors. To propose that a chimp is capable of mental images is actually sensible, in view of how chimps respond to words, and what they can recall about real objects merely from words, a topic which will come up a bit later.

The color of the plastic shapes had no resemblance to the colors they named, of course. Every effort was made to eliminate iconic representation in the shapes of the plastic as well. Only Sarah's name, a silhouette of a monkey, could be considered iconic, though Sarah would probably disagree.

Sarah learned "all," "one," "none," and "several," using round and square crackers. "One" described a single round cracker among four square ones, or the reverse, one square and four round. Two or three round crackers described "several," or the reverse. In these tests, after Sarah learned the quantifiers, instead of being asked questions, she was told:

Sarah insert all cracker dish	Sarah insert one cracker dish	Sarah insert several candy dish

and so on. She was allowed to keep for herself the crackers or candy placed in the dish. And most of the time she carried out her instructions rather smoothly, with one exception. When asked to place "one" cracker in the dish during a lesson, Sarah managed to sneak a few candies in as well. Mary Morgan decided to overlook this endearing bit of cheating, mainly because the sentence did not actually negate such an act. And, as far as the crackers were concerned, she had placed only the one cracker in the dish. But when Sarah read:

```
Sarah
insert
one
candy
dish
```

she terminated the lesson by simply grabbing all the candy. Had she placed just what she was told to place in the dish, she would have received just one piece of candy. Sarah opted for the whole bag.

7

understanding

sentences

In the wild, a young chimpanzee watches a variety of scenes: its mother eats a fruit, she hands a tidbit to his infant sister, then hands the little sister to another chimp while she joins a hunting group. Sarah observes her teachers bring a basket of goodies to the laboratory, she takes and plays with toys, returns them, nibbles on raisins and fresh fruit. We, the wild chimp, and Sarah see entire actions at a time since our visual image of a scene is compressed like a chord. But every scene is an incipient sentence. Every scene can be translated into an actor and action.

When we describe a scene with a sentence, we take the visual chord apart, separate it into notes, and string the notes out separately: "Mother eats pawpaw," "Mother gives meat to sister," "Teachers bring toys to Sarah." The chord translates into notes quite as the visual image translates into words, which are strung out in a linear fashion. Actors, actions, objects, indirect objects, and so on are the notes out of which sentences are written. It is how we divide scenes into sentences. Can chimps divide the world in the same way? We know from Nadezhda Kohts's research that chimps have the same color and form vision as we do. But, to translate a scene into a sentence, chimps must be able to use symbols to name actors, actions, and so on. Chimps show good evidence for being able to symbolize: the chimps in Wolfe's study used poker chips as symbols for food and actions, and chimps taught to sign have learned over one hundred gestures for an even wider range of categories.

Signing chimps even use sequences of signs: "Please gimme sweet, tickle Roger Washoe, gimme gimmme gimmme sweet." But, to understand sentences, chimps must do more. They must use words in sequences and follow rules that will produce an orderly sentence. It is only because we order our words according to commonly understood rules that we can communicate with one another using language. Before Sarah's education began, there was no evidence a chimp could integrate the use of symbols with rules of sentence order.

Ordering a simple sentence

Sarah's words were made of plastic, with metal backing, dyed in several colors, cut in a variety of shapes, and stored in her dictionary, a huge but shallow container with a separate bin for each word. Before the daily lesson, the trainer stopped at the dictionary and picked up a bucket of vocabulary for the hour-long class. Every trainer had a name as well: a peace sign, a large letter M, a star of David, and so on, which were worn as necklaces. At first, Sarah could not understand how to use her words, but one of the trainers, now Dr. James Olson, entered Sarah's cage to demonstrate and she soon was intrigued by the magnetization.

If a plastic word slithered off the board, as it did on occasion when the magnetic hold was weak, Sarah retrieved the word and immediately returned it to the board, trying carefully to prevent it from sliding down again. She soon developed a special attitude toward her words, treating them with care, and later, when her sentences became long, she used her mouth as a small bucket in which to hold them before returning them to the table. She wrote sentences vertically, her preference, and the trainers accommodated themselves to her, since Sarah cooperated by learning and ordering words.

Sarah and her trainer sat across from one another while the trainer placed a piece of fruit on the table and looked on indulgently as Sarah picked up and ate the fruit. After a time, the trainer introduced a plastic word along with the piece of fruit; the fruit was not as easily available now as the plastic word. Sarah was coaxed to place a word on the board, which she did, and promptly got the piece of fruit. After Sarah ate her fruit slice, she removed the word from her board and returned it to the trainer. Soon she received two fruit names but only one piece of fruit, and now, to have the fruit, she had to name it correctly.

Soon she was given two pieces of fruit and two fruit names, and to receive an apple slice Sarah had to place the word "apple" on the board. When she knew the two fruits, banana and apple, she learned the names of her trainers, Mary and Randy, both of whom wore their names as pendants around their necks. To receive a fruit, Sarah had to identify each trainer by name, but now, since sentence order was introduced, she was required to write:

Mary apple	Randy apple	Randy banana	Mary banana

in order to be given the fruit. Sarah next wore her own

name as a necklace and had to name herself as the receiver in all the actions of giving: "Mary apple Sarah," "Randy banana Sarah" and so on. Because Sarah never gave anything to her trainers she could not be taught to distinguish receivers—she was the only one. To remedy this limitation, Gussie's name appeared on Sarah's table one day, and when, in all innocence, Sarah placed the name on the board, the apple was given to Gussie, Sarah's neighbor. Sarah had a tantrum and refused to continue or place the name "Gussie" on her board again.

As an inducement to higher generosity, when Sarah was coaxed to place Gussie's name on the board once more, though Gussie did receive the apple again, Sarah was given something she much preferred, a piece of chocolate. Now Sarah could write: "Mary banana Gussie, Randy apple Sarah, Mary apple Gussie," and so on. She not only described each scene with words but used the words in correct order as well.

Finally Sarah learned that actions could vary by observing scenes in which various fruits were cut, washed, and inserted into containers, as well as given. When donors increased in numbers, actions varied more widely, objects could be eaten or not, and receivers multiplied, Sarah could both understand and write such sentences as:

Mary insert apple	Jim give ball Sarah	Randy wash banana	Sarah cut banana	Mary give apple Gussie

In these sentences, the actual scenes were not easily

reversed. For instance, apple could not insert Mary, nor was Sarah likely to give things to others. Order was essentially built into the semantics of the sentence: only animate creatures can wash, cut, give. Action moves in one direction only. But in the prepositional world there are no such limitations. Virtually anything can be "on" something else, and both scene and sentence are reversible. Sarah was now taught to observe a situation in which colored cards were set "on" cards of a different color. Green cards were sometimes on blue ones, or blue ones on green, reds on yellows, or blues on reds. Any card could be on another card, one situation could easily be reversed, and Sarah learned to describe these changing scenes by writing:

red on green	yellow on red	green on red	green on blue

in the appropriate situations. On some occasions Sarah was asked:

? red on green	Is red on green?

If, in actuality, *green* was on red, instead of replying "no," on at least 30 percent of the trials Sarah changed the position of the cards, placing red on green, and answered "Yes." Her general dislike for the negative inclined her to change the world to agree with her language preference for "yes."

Forming a compound sentence

It was slightly more difficult to teach Sarah a compound sentence. By dropping the repetition of certain words in "Sarah give apple Mary," "Sarah give banana Mary," simple sentences become "Sarah give apple and banana Mary." Although "and banana Mary" seems to be an unrelated string of words, we understand it is Sarah who gives. But the ability to organize language information in this way is not intuitive; it must be learned. There was no reason to believe Sarah could understand that "and banana Mary," a seemingly unrelated string of words, was an integral part of the sentence.

Sarah faced a dish and a sand pail, two fruits, an apple and a banana, and when shown the sentences: "Sarah insert banana pail," or "Sarah insert apple pail," or "Sarah insert banana dish," or "Sarah insert apple dish," she followed each instruction. When it was clear she could read and interpret the simple sentences, she was shown a pair of sentences at a time,

Sara	Sara
insert	insert
apple	banana
pail	dish

She followed the directions of both,

then finally responded correctly to a compound sentence.

```
Sarah
insert
apple
pail
banana
dish
```

We thought she might be confused by the sentence, might place the banana in the pail, for example, or simply ignore banana and dish, which seemed to have little relation to the subject and verb. But Sarah organized the sentence accurately, and performed both actions correctly. She also understood other forms of compound sentences, carrying out their directions accurately.

From simple to complex sentences

Sarah was next introduced to the conditional sentence, "If Sarah good then Mary give Sarah kiss," for example. In this complex sentence, David decided against using discontinuous elements; instead, Sarah was taught the sign $\supset$, which in logic symbolizes "if-then." The sentences were now stated: "Sarah is good if-then Mary give Sarah kiss." To give Sarah actual experience in a conditional situation, she received contingency training, which is almost a model of the conditional sentence. In the usual contingency situations an animal is shown two pieces of fruit and is free to take one or the other. When it takes the slice of apple it receives a piece of chocolate, but when it takes a slice of banana it is given nothing. This type of contingency experience is also called reinforcement training; if the animal is fond of chocolate, it will soon choose the apple most frequently. Sarah, also fond of chocolate, chose the apple most frequently during the contingency experience. She soon stopped taking the slice of banana, for she never was given anything for choosing that particular fruit. Sarah's experience was translated into two sentences which described the contingency situation in the form of a question:

Sarah
take
apple
?
Mary
give
Sarah
chocolate

"What is the relation between Sarah's taking apple and Mary's giving her chocolate?" Sarah replaced the interrogative element with her only word, the conditional sign $\supset$. Sarah was given a slice of apple followed by a piece of chocolate. Then came the next question, "What is the relation between Mary's not giving Sarah chocolate and Sarah's taking banana?" and the $\supset$ again replaced the ?. Sarah was given a slice of banana and that was all.

Next Sarah had a slice of apple and a slice of banana before her, while on the board Mary arranged a set of sentences:

Sarah	Sarah
take	take
apple	banana
if then	if then
Mary	Mary
give	not give
Sarah	Sarah
chocolate	chocolate

Or she would arrange the opposite, that is, sentences in-dicating Sarah would receive chocolate if she took the banana, not the apple. Since Sarah was so fond of chocolate she was considered to have understood cor-rectly if she selected the fruit that promised chocolate. But Sarah seemed to understand nothing—she chose the ap-ple on every test, whether it promised chocolate or not, and she shrieked in fury over her errors. Unfortunately, her contingency training experience had involved the con-sistent reinforcement of the choice of apple, not banana. Faced with a difficult lesson, Sarah reverted to her earlier learning pattern rather than read her instructions. She soon discovered this error was the source of all her frus-tration. She then began to pay close attention to the sen-tences on her board, reading them with care, and subse-quently she performed the rest of her problems correctly. Once she learned the conditional relationship, she was able to handle her rather difficult transfer tests with con-siderable ease, as easily, in fact, as she had handled the compounds.

In this series, changes appeared in both the antecedent and consequent portions of the sentences, for instance,

"Mary take red if then Sarah take apple: Mary take green if then Sarah take banana." Sarah had to watch what Mary did in order to perform the correct action herself. If Mary took a red card, Sarah was to take the apple, and if the green one, she was to take the banana. In addition, with these sentences

red	green
on	on
green	red
if then	if then
Sarah	Sarah
take	take
apple	banana

which reviewed Sarah on her prepositional sentences, she performed very well.

Sarah learned the concept of a relationship first with simple objects such as cups and spoons. When she began to construct sentences on the board, she learned there were relationships between words in simple sentences, and finally that relationships also existed between simple sentences. From this point she moved to the compound and finally comprehended the dependency relation within complex sentences.

Clever Hans

Many years ago, a man in Germany trained his horse to perform the most astounding feats in response to verbal instructions. Then it was discovered that the horse, Hans, was responding not to the verbal instructions, but to subtle cues by which his owner provided the answers. Of course, the owner was quite innocent of deliberately providing such cues, but he was discredited and his reputation lost. His specter is resurrected, however, when animals begin to do more than they properly should.

In the research with Sarah, everyone was particularly cautious about giving cues. It was desirable, nevertheless, to check on the possibility Sarah was learning cues rather than language, so she was adapted to a new trainer, John. Though he interacted in a friendly way with Sarah, he was never taught Sarah's language. He was given a name to wear and recognized his own name as well as Sarah's but did not have any further knowledge of Sarah's language.

Two trainers were involved in this test—one sophisticated, the other naïve. John remained in the test situation with Sarah, while the informed trainer remained out in the hall. Each plastic word was coded by number on John's sheet of paper, from which he followed his numerical directions; then he reported Sarah's choices, by number, over a microphone to the informed trainer in the hall. If Sarah performed correctly on a trial, John was told "yes" through the headphones and he gave Sarah fruit and encouragement. When John heard "no" he gave neither fruit nor praise. On all of these tests, Sarah performed less well than on her usual lessons, where she was about a B student. But she tested well enough to counter arguments she might be another clever Hans.

Besides performing less well with the new and uninformed trainer, Sarah reverted to earlier sentence production habits. She had been placing words in final order in the first attempt, but now she returned to her old habit of placing her words on the board in scrambled order, then tidying up. Also, her nice vertical line was replaced by

a skewed one, another sign of deterioration in her pen-
manship. In spite of the emotional disruption, faced with
her new trainer, deprived for a few days of her familiar
faces, Sarah still performed well above chance. It would
be interesting to see how well a child at this language
stage of about 150 words would do in a simple language
test with a virtual stranger. Certainly, if the child did not
perform as well as usual, we would not say the child was
incapable of language.

What does a word mean?

When we use words for objects and events, we cannot
be sure if we use these words in the same way others do.
Children say "doggy" to almost any animate creature provided it has four legs, and "mama" is the name for a number of people who take care of the child. The chimpanzee
Washoe used the sign for flower in the presence of
tobacco. The salient feature of the flower, for Washoe, was
its odor and the salient feature of the dog is its four-
leggedness. So Washoe signs "flower" to aromatic
things and the child says "doggy" to goats and cats. It
was only reasonable to assume Sarah did not use words
in the same way as do sophisticated speakers of English.

Sarah used her words in the proper contexts, but since
she is a chimp and not human, David wondered what the
colored pieces of plastic actually meant to her. When we
use words we don't really think of them as words. They
have become so ingrained in our thinking, they *are* what
they name. If I showed an older child a picture of her
father and asked her to describe the picture, she might
say, "Oh, he has brown eyes, big ears, lots of teeth, a
beard, etc." If I later asked her to describe "daddy," she
would probably repeat, ". . . brown eyes, big ears, lots of

teeth. . . ." She definitely would not say, "Oh yes, daddy, a five-letter word, has three d's, rhymes with baddy. . . ." When asked to describe a word, a human describes what the word stands for, not the actual word itself.

The possibility that Sarah might regard her words in the same fashion was intriguing. David designed a simple test to analyze Sarah's view of words. She was given an apple, and asked to choose between sets of features that best described an apple from her point of view:

round	vs.	square
red	vs.	green
round	vs.	round with stem
square	vs.	square with stem
round	vs.	square with stem

After Sarah had chosen the features she considered most like those of the real apple, she was given a test with her *word* for apple. Once again Sarah chose the features round, red, round with stem for an actual apple, but in the presence of her *word* for apple, a light-blue plastic triangle!

Sarah's language education began when she was six and continued until she was about ten. During this period she was taught a large number of language exemplars. Early in her training she learned to differentiate "same" and "different" and was taught the question. She learned color, shape, size, and knew several instances of each class, i.e., red and green were colors, round and square were shapes. The quantifiers, "all, none, several, one," became part of her repertoire as well. Besides learning labels for objects and actions, Sarah "described" visual

"Once again Sarah chose . . ."

scenes with sentences. She produced and comprehended a wide variety of simple sentences and could follow the directions of both compound and complex sentences, indicating she understood something of sentence structure. Sarah's linguistic performance is difficult to categorize. Washoe learned as many words as Sarah and both used words in sequences, though Washoe's signs did not follow sequential rules. Dr. Duane Rumbaugh at Yerkes in Atlanta, Georgia (no relation to the original Yerkes), is teaching his chimp Lana a language based essentially on Sarah's visual system. Lana uses a custom-built typewriter with symbols rather than letters on the keys. She follows a few word sequences, as "please machine give apple period," and if correct, receives the fruit. If her sequence is incorrect, she can cancel it and try again. Lana learned several sequences by the age of three, after six months of training.

Dr. Roberta Klatzky, a psycholinguist at the University of

California, says that at the two-word grammar stage children know a few colors, the word "big," and a few other adjectives. By the time the child is three years old, it has mastered the specific properties "big" and "small." Other studies suggest children know shapes, colors, and sizes quite reliably at the age of five. Children between the ages of two and three can handle both negatives and questions: "No go car," "Where Daddy?" And, although they comprehend some clauses, they don't seem to be able to handle the conditional sentence before the age of five. This places Sarah somewhere within the first five years of human performance in the particular language exemplars she learned, except for the conditional sentence, which children under five have not yet learned. Washoe performed at about the two-and-one-half-year-old level, while Lana has only begun to demonstrate her ability.

It is clear, from the knowledge of chimpanzees we have to date, that they can be taught a simple language. What is done with this knowledge in the future will depend on those who continue to work with chimps. An adventurous person might travel to the forest to teach language to "wild chimps," bringing a staff of trained chimps or people to teach language to chimps in the wild. We would undoubtedly learn much more about chimp culture through a language exchange rather than by observation alone. But language knowledge gleaned from chimps can have important applications for people as well.

Speech is not the only avenue for language. There is little doubt children make cognitive distinctions long before they are able to speak. Deaf children, for instance, learn their first signs months earlier than hearing children say their first word, as manual dexterity develops much before vocal dexterity. Mary Morgan, Sarah's main

Sarah's Partial Vocabulary

Proper nouns	Nouns— object names	Particles
Jim	dish	yes
Mary	cup	no
Randy	key	
Sarah	shoe	*Conjunction*
Ann	spoon	if then
Gussie	paper	
John	bubble	*Others*
	flashlight	color of
Nouns—foods	sponge	shape of
apple	bottle	size of
apricot	soap	name of
banana	toothbrush	
cherry	comb	
fig		
orange	*Adjectives*	
grape	same	
raisin	different	
caramel	blue	
chocolate	green	
crackerjack	red	
gumdrop	orange	
	yellow	
	brown	
	square	
	round	
	small	
	big	

trainer, taught her baby of ten months to identify several baby foods with colored blocks. At the time, the baby could say "mama" only, but she could, given a sniff of dinner, name it by pushing the correct block into a small chute built into her high chair. If she identified the food correctly, she could eat it. Even a baby can label items if we adjust the labeling system to its ability.

It is clear that thousands of children *cannot* acquire language simply by living in a speaking environment. But the situation is not without hope. "If you can teach a monkey to talk, why can't you help my child?" the mother of a speechless child once asked David. The question is legitimate and one which is now being answered. The procedure for teaching chimps has been used with considerable success among language-deficient children. Because chimps have learned to read, children can be taught language in a novel and effective way.

bibliography

Chapter 1

Garner, R. L. *Gorillas and Chimpanzees.* Osgood, McIlvaine and Co., 1896.

Goodall, Jane. "Chimpanzees of the Gombe Stream Reserve." In Irven Devore, ed. *Primate Behavior.* Holt, Rinehart & Winston, New York, 1965.

Kortlandt, Adriaan. "Chimpanzees in the Wild." *Scientific American* 206 (5) (1962): 128–138.

Nissen, H. W. "A Field Study of the Chimpanzee." *Comparative Psychology Monographs* 8, no. 1 (Serial no. 36), 1931.

Reynolds, Vernon and Frances. "Chimpanzees of the Budongo Forest." In Irven Devore, ed. *Primate Behavior.* Holt, Rinehart & Winston, New York, 1965.

Struhsaker, Thomas T. "Auditory Communication Among the Vervet Monkeys." In Stuart A. Altmann, ed. *Social Communication Among Primates.* University of Chicago Press, Chicago, 1967.

Van Lawick–Goodall, Jane. *In the Shadow of Man.* Houghton Mifflin, Boston, 1971.

Van Lawick–Goodall, Jane. "Mother-Offspring Relationship in Free-Ranging Chimpanzees." In Desmond Morris. *Primate Ethology.* Aldine Publishing, Chicago, 1967.

Yerkes, Robert M., and Ada W. *The Great Apes.* Yale University Press, New Haven, Conn., 1945.

Chapter 2

Gallup, G. G., Jr. "Chimpanzees: Self-recognition." *Science* 167 (1970): 86–87.

Hayes, Cathy. *The Ape in Our House.* Harper & Row, New York, 1951.

Hayes, D. J., and C. H. Nissen. "Higher Mental Functions of a Home-Raised Chimpanzee." In A. M. Schrier and F. Stollnitz, eds. *Behavior of Nonhuman Primates* (Modern Research Trends), vol. 4. Academic Press, New York, 1971.

Jacobsen, F. Carlyle, Marian M., and Joseph G. Yoshioka. "Development of an Infant Chimpanzee During Her First Year." *Comparative Psychology Monographs* 9, no. 1 (Serial no. 41), 1932.

Kellogg, W. N. *The Ape and the Child.* McGraw-Hill, New York, 1933.

Kohts, N. *Infant Ape and Human Child.* Scientific Memoirs of the Museum Darwinianum in Moscow, 1935.

Yerkes, Robert N., and Alexander Petrunkevitch. "Studies of Chimpanzee Vision by N. Ladygina-Kohts," trans. from the Russian. *Journal of Comparative Psychology* 5 (1925): 99–108.

Chapter 3

Kawai, M. "Newly Acquired Precultural Behavior of the Natural Troop of Japanese Monkeys on Koshima Islet." *Primates* 6 (1965): 1–30.

Köhler, Wolfgang. *The Mentality of Apes.* Humanities Press, New York, 1951.

Kummer, Hans. *Primate Societies.* Aldine Atherton, Chicago, 1971.

Teleki, Geza. "The Omnivorous Chimpanzee." *Scientific American,* January, 1973.

Chapter 4

Menzel, Emil W. "A Locational Approach with Applications to Chimpanzee Group Psychology." In A. M. Schrier and F. Stollnitz, eds. *Behavior of Nonhuman Primates,* vol. 5. Academic Press, New York, in press.

Miller, Robert E. "Experimental Approaches to Affective Communication." In Stuart A. Altmann, ed. *Social Communication Among Primates.* University of Chicago Press, Chicago, 1967.

Ploog, Detlev W. "The Behavior of Squirrel Monkeys as Revealed by Sociometry, Bioacoustics, and Brain Stimulation." In Stuart A. Altmann, ed. *Social Communication Among Primates.* University of Chicago Press, Chicago, 1967.

Suomi, Stephen J., Harry F. Harlow, and William T. Mckinney, Jr. "Monkey Psychiatrists." *American Journal of Psychiatry* 128 (February 1972): 8.

Wolfe, John B. "Effectiveness of Token Rewards for Chimpanzees." *Comparative Psychology Monographs* 12, no. 5 (May 1936).

Chapter 5

Bellugi, Ursula, and Edward S. Klima. "The Signs of Language in Child and Chimpanzee." In T. Alloway, et al., eds. *Communication and Affect*. Academic Press, New York, 1972.

Bellugi, Ursula, and Edward S. Klima. "The Roots of Language in the Sign Talk of the Deaf." *Psychology Today,* June, 1972.

Braine, M. D. S. "The Acquisition of Language in Infant and Child." In C. Reed, ed. *The Learning of Language.* Appleton-Century-Crofts, in press.

Gardner, R. Allen, and Beatrice T. "Teaching Sign Language to a Chimpanzee." *Science* 165: 664–672.

Gardner, R. Allen, and Beatrice T. "Two-Way Communication with an Infant Chimpanzee." In A. M. Schrier and F. Stollnitz, eds. *Behavior in Nonhuman Primates,* vol. 4. Academic Press, New York, 1971.

Chapters 6 and 7

Glass, Andrea Velletri, Michael S. Gazzaniga, and David Premack. "Artificial Language Training in Global Aphasics." *Neuropsychologia* 11 (1973): 95–103.

Itard, J. M. G. *The Wild Boy of Aveyron* (trans. G. M. Humphrey). Century, New York, 1932.

Pfungst, O. *Clever Hans, the Horse of Mr. van Osten.* Holt, Rinehart & Winston, New York, 1911.

Premack, Ann James, and David. "Teaching Language to an Ape." *Scientific American,* October, 1972.

Premack, David. *Intelligence in Ape and Human.* Lawrence Erlbaum Associates, New Jersey, in press.

Premack, David. "Language in Chimpanzee?" *Science* 172 (May 21, 1971): 808–822.

Premack, David. "On the Assessment of Language Competence in the Chimpanzee." In A. M. Schrier and F. Stollnitz, eds. *Behavior of Nonhuman Primates.* Academic Press, New York, 1971.

Rumbaugh, D. M., T. V. Gill, and E. C. von Glaserfield. "Reading and Sentence Completion by a Chimpanzee (Pan)." *Science* 182 (November 16, 1973): 731–733.

index